The Origin and Cultivation of
Shade and Ornamental Trees

THE MORRIS ARBORETUM

The Origin and Cultivation
of
Shade and Ornamental Trees

by Hui-Lin Li

Philadelphia
University of Pennsylvania Press

Preface

This book is a treatise on the historical aspects of our ornamental and shade trees. Its purpose is to provide information about the origin, beginning of cultivation, and variation under domestication of the common species of trees that are closely associated with men in their settled habitations. The trees herein discussed belong primarily to the streets, lawns, gardens, and parks of the temperate world. Cultivated trees of tropical origin are distinct botanically as well as culturally from trees of the temperate world, and thus they cannot be included in the same treatment.

In tracing the origins—historical, geographical, botanical, and horticultural—of trees, the subject touches various phases of human activities. It is not a subject that can be confined to the trees alone, and therefore is not one of concern only to the professional horticulturists, arborists, and dendrologists, but one, at least in some of its aspects, that is also of interest to other students, such as historians, geographers, and anthropologists. Even the home gardener or the nature lover may be interested in gaining some background about the trees in their very immediate surroundings. To reach this wider circle, the technical terminology in the following discourse is kept at a minimum. While this book has been written primarily for the general reader, a bibliography is provided for those who wish to have access to sources for further information.

Tree cultivation began in the dim past of human civilization and some of our most important trees are thus of mythical origins. Others were introduced into cultivation in historical times and the origins of many of these can still be documented. Important epochs of tree culture in more recent times include the opening up of the New World and the penetration of the interior of Asia for botanical exploration. An attempt has been made to present a chronicle of tree culture by tracing from the most ancient trees down to the latest, most outstanding case of introduction, the *Metasequoia*. Classified enumerations are also given to present the geographical aspects of the subject.

Chapters I to IX are based on material originally published as articles in the *Morris Arboretum Bulletin*, 1956-1960. Thanks are due to Dr. John M. Fogg, Jr., Director of the Morris Arboretum, for his help in the original publications and for his permission to use the material here. The sources of some of the photographs were originally acknowledged in the Bulletin and credit is also given here along with the individual figures. Those illustrations that are without a credit line are either photographs or drawings taken or prepared by myself.

*To the
Memory of My Mother*

CONTENTS

The Origin and Cultivation of
Shade and Ornamental Trees

The Tree Planter

Whoever planted rows of trees
Beside the roads and lanes
God rest his soul in Heavenly peace
And bless him for his pains;
For he who gave of time and toil,
Who gave of heart and hand
To nurse the tender shoots that were
To shade of ways of man,
Was quite as great as those who built
Of stone and minted gold—
No need to cast his name in bronze,
His deeds need not be told.

—*Stanley Foss Bartlett*

1

The Cultivation of Trees
by Mankind

From the very beginning of the human race, man and trees have been intimately associated. Trees, as civilization gradually progressed, became increasingly important to mankind by providing such basic needs as shelter, clothing, and food, as well as drugs, poisons, and dyes. Consequently trees played a not insignificant role in the religion of primitive peoples and were one of the most frequently mentioned subjects in early legends and folklore.

Man's earliest associations with trees were, of course, with those of the native forests. At a very early date, however, some trees were deliberately planted, long before any written languages began to appear to record such an event. Most probably, the earliest plantings of trees happened at many different times and places independent of each other.

BEGINNING OF AGRICULTURE

As the ice sheets of the last Glacial Age began to recede, some 50,000 years ago, man, the primeval savage, entered slowly into what is now called the Paleolithic Age. He was then a hunter of food and he was both carnivorous and herbivorous. Besides whatever animals he could capture with his crude stone implements, all kinds of nuts, berries, fruits, herbs, and succulent rootstocks were in his diet. Plants probably furnished more of his food than animals, and he had apparently acquired some knowledge about plants from his remote aboreal ancestors.

It is generally believed that the present races of mankind appeared some 15,000 years ago, and this marks the dawn of civilization. This is known as the Neolithic Age, as man was able to make more polished stone implements, to produce pottery, and to begin domesticating animals and cultivating plants. He was able also to utilize plant materials for plaiting and weaving.

All these changes that took place were, of course, very gradual and there is no sharp break between one phase of culture and the other. Just how and where man first acquired the knowledge of cultivating plants and utilizing seeds is a story lost in remote antiquity that can never be clearly revealed to us.

We know, however, the Neolithic man in Europe cultivated wheat, barley, and millet. He used crude implements made of wood for cultivation. He also ate peas and crabapples and perhaps also cultivated these in their wild forms.

BEGINNING OF TREE CULTIVATION

The cultivation of certain fruit trees might have begun as early as the cultivation of cereals or perhaps even before. This is, however, purely a hypothetical suggestion. Primeval man apparently used wild nuts, acorns, and fruits before grain cereals. However, the knowledge of growing plants from seed would be more readily acquired from annual herbs than from trees with a much longer life span.

Although it might be difficult for the primitive man to comprehend the relationship between planting a seed and the kind of mature tree eventually grown from it, it should be noted that some trees grow easily from cuttings, a process that can be readily perceived. The use of stakes for making protective hedges and of timbers for building shelters and the accidental rooting of some of the materials would soon reveal to the early man the secret of growing and propagating certain trees. Moreover, the early development of weaving and plaiting with such materials as willow twigs, which grow readily from cuttings, would similarly contribute to such an advancement in human knowledge.

Whatever scattered knowledge on growing plants developed among the primitive peoples could be brought to others by migrations, warfare, and other intermixtures. It is the diffusion and accumulation of knowledge that brought about continuous advancement in civilization.

Between some ten to twenty thousand years ago, as primitive man developed agriculture and gradually settled down to food producing instead of food hunting, a great change occurred in the living conditions of mankind. Settlement began in small communities, which made possible rapid development of various phases of human civilization. With these settlements, trees were planted to afford shade and protection, to provide edible fruits and nuts, and to furnish symbols for worship and as memorials. Naturally grown trees of great size or age or of high yield in fruit were probably first singled out for preservation, but later trees were deliberately planted for these purposes.

The first recorded experiment in shade-tree gardening is discovered from a Sumerian tablet of about the third millenium B.C. It is inscribed with a myth, made known for the first time by Prof. S. N. Kramer in 1946 (Kramer 1956), concerning the rape of the goddess Inanna by the gardener

Shukalituda and noted for a "blood-plague" motif similar to that of the Biblical exodus story. The myth begins with an horticultural experiment undertaken by the gardener. After continuous failure at his gardening, Shukalituda acquired new wisdom and planted the *sarbatu* tree (as yet unidentified) in the garden, a tree whose broad shade lasts from sunrise to sunset. As a consequence of this experiment, his garden blossomed forth with all kinds of greens. Kramer thinks the author of this ancient myth "seems to explain the origin of shade-tree gardening, and thus reveals that the horticultural technique of planting shade trees in a garden or grove to protect the plants from wind and sun was known and practiced thousands of years ago."

THE TREE OF LIFE

Trees were frequently mentioned in the Bible. The second chapter of Genesis says, "And out of the ground made the Lord God to grow every tree that is pleasant to the sight, and good for food; the tree of life also in the midst of the garden, and the tree of knowledge of good and evil."

The tree of life conferred on man immortality. The tree of knowledge (Figure 1) gave the power of distinguishing good and evil. One was moral and the other prophetic, the sign of the first revelation to man. They are in Paradise and were intended to teach the primitive man in moral duty, and in his anticipation of the world to come.

We are now obliged to connect the early chapters of Genesis with the old beliefs of Babylonia. In another ancient civilization farther east, in China, the tree of life can also be traced in their ancient traditions. In the writings of the philosopher Li-tze, who lived about 450 B.C., there is

1. Many versions of the "tree of knowledge" are depicted in publications of all ages. This one is from the herbal, Ortus Sanitatis, 1491.

mention of the Fairy Islands in the eastern ocean, a paradise of gold and jade palaces, beauteous birds, and trees whose fruits confer immortality (Edkins, 1889).

TREES FOR WORSHIP AND AS MEMORIALS

Trees are thus associated with the earliest religious beliefs of man. In ancient times among all peoples trees of massive size and great age became mystical symbols and objects of worship. In the mythology of all races, trees are very frequently mentioned. Even to this day old trees are often still worshiped or revered among primitive or superstitious peoples.

In the Scriptures trees are mentioned not only for use as ornamentals but were also planted, as is still the case in cemeteries, as memorials. Later on in Roman days, Pliny observes, "In old times trees were the very temple of the gods; and, according to that ancient manner, the plain and simple peasants of the country, savouring still of antiquity, do at this day consecrate to one god or other the godliest and fairest trees that they can meet withall; and verily, we ourselves adore, not with more reverence and devotion, the stately images of gods within our temples, the very groves and tufts of trees, wherein we worship the same gods in religious silence" (Holland's translation of Pliny's *Natural History*, p. 357.)

Pliny goes on to say, "The ancient ceremony of dedicating this and that kind of trees to several gods, as proper and peculiar to them, was always observed, and continuous to this day. For the mighty oak, named esculus, is consecrated to Jupiter, the laurel to Apollo, the olive to Minerva, the myrtle to Venus, and the poplar to Hercules."

In the Far East the use of trees as memorials also began very early. In the Chou dynasty (1122-240 B.C.) long traditions had already established the five official memorial trees for the tombs: pine for kings, arborvitae for princes, sophora (pagoda tree or Chinese scholar tree) for higher officials, koelreuteria (China tree) for scholars, and poplars for the common people (Figures 2-5).

2. *An ancient juniper planted in Central Park, former imperial garden, Peking, China. (From J. Arnold Aboretum, 1926.)*

3. Old trees of arborvitae in Central Park, Peking, China (From J. Arnold Arboretum, 1930.)

4. Cryptomerias on temple grounds, Nikko, Japan. (From Elwes & Henry, Trees of Great Britain & Ireland, 1906.)

5. A huge yew tree in a church yard, Tisbury, England. (From Elwes & Henry, Trees of Great Britain & Ireland, 1906.)

FRUIT AND NUT TREES

The origin of cultivation of many of our common fruit trees is lost in antiquity. Their domestication began in prehistoric times and thus it is now impossible to trace their ancestral forms. In some cases the wild ancestral species are probably still extant, but they bear only remote resemblance to the cultivated types which have changed greatly under long cultivation. In most others the original wild types are long extinct. There may be occasional claims of discoveries of the wild types of certain of our long-domesticated fruit trees, but these are often only naturalized escapes of culti-vated plants and not spontaneous wild forms.

The absence of wild prototypes of some of our common fruit trees, like many of the cereals, attests to their great antiquity. Such fruits as apples, plums, pears, peaches, and cherries are among the earliest cultivated plants of man. These fruit trees originated in two major centers, cor-responding to the two great centers of civilization of the Northern Hemisphere: Eurasia, that is, western Asia and Europe, and eastern Asia. Among the different apples, pears, plums, and cherries, there are Eurasian groups as well as Chinese groups of each kind. There are also grape, fig, date palm, and others of western Asiatic origin, and peach, oranges and others of eastern Asiatic origin only.

Some general ideas about the process of early cultivation and domestication of these fruit trees, long lost in antiquity, can be gained by observing the development of fruits in the New World in recent history. In temperate North America, there are many wild species of *Prunus* and *Vac-cinium* and other genera that bear edible fruits. The in-troduction of these trees into cultivation began only in the last one to two hundred years. Of the many species, a few have proved to be promising fruit trees, such as plums, blueberries, and cranberries, but others, such as native apples, cherries, and other berries, have been found to be of little or no value. Thus, although many species are avail-able and tried out at first, most fall into disfavor in one way or the other. Gradual elimination will eventually leave only a few of the most worthy ones. These, by intense selection and improvement, will ultimately become so different from their wild prototypes that their relationships may be diffi-cult to recognize by future generations. The general course of development of fruit trees in the Old World in pre-historic times is undoubtedly similar.

6. *Walnut in Barrington Park, England. (From Elwes & Henry, Trees of Great Britain & Ireland, 1906.)*

Acorns and nuts must have been used by man since the remotest times. The remains of the Peking man, living some 100,000 years ago in northern China, show that hackberries were already used as food. The improvement of nut trees under cultivation, however, has not been carried out as far as the edible fleshy fruit ones.

The most valuable and widely used nut of the whole world is the coconut, a maritime palm of the tropics. There are many other nut trees in the tropical and subtropical regions of the world. The most important nut of the temperate world is the walnut, *Juglans regia*, (Figure 6) a native of central Asia to southeastern Europe. Other important nut trees of the North Temperate Zone are chestnuts, hazelnuts and almond.

For further details on the origins of the various fruit and nut trees, the interested reader can consult such authoritative treatments by DeCandolle (1884) and Vavilov (1926, 1951).

TREES FOR SHADE AND ORNAMENT

In nearly all cases the ancestral forms of cultivated shade and ornamental trees are still known. This shows that the planting of trees for shade and ornament began later than fruit trees. These trees, as compared with fruit trees, are less variable, indicating also their relatively shorter history. Furthermore, the parentage of most of the hybrid trees originated in cultivation can still be clearly traced.

In the Scriptures, trees planted for ornament are frequently mentioned. Solomon, for instance, transplanted cedar to the plains from the mountains. The cedar is, in the Book of Ezekiel, to be frequent in magnificent gardens.

Reliable information on trees known to the ancients, down to the time of the Greeks, is to be found in the works of Theophrastus. Many of the plants mentioned by him were identified by Sprengel (1808). Sprengel's identifications were subsequently revised and amended by Stackhouse in his edition of Theophrastus' *Historia Plantarum*. A list of not less than 170 ligneous plants selected out of this work is given by Loudon (1838). These include many trees and shrubs native to Greece and also others such as peach,

7. *A yew avenue at Midhurst, England. (From Elwes & Henry, Trees of Great Britain & Ireland, 1906.)*

persimmon, and cherry which were introduced from abroad. While most of the plants are economic ones, a number of the trees valued mainly for shade or ornament such as elm, plane, yew, beech, and alder are also listed (Figures 7,8).

The writings of the Romans show their knowledge of all the trees possessed by the Greeks and also the trees of

*8. An old beech tree at Newbattle, England. (From Elwes &
Henry, Trees Gt. Brit. & Irel., 1906.)*

the colder regions of Europe. Among ornamental trees, the
pine, the bay, and the box appear to have been favorite
trees in gardens. Many other trees were planted for various
useful purposes, such as for their wood and fruit, or as
fuel and other usages. The most reliable source of information on trees of the Romans may be found in Pliny's *Natural
History*. The species mentioned there were also identified
by Sprengel.

In ancient China, the highways of the Chou dynasty

(1122-240 B.C.) were famed for their smoothness. These highways were known to be lined with trees. During the short-lived Ch'in dynasty (221-206 B.C.), China was unified for the first time as a single empire. The Emperor, Shih Huang Ti, had military superhighways built from the capital for thousands of miles in all directions to the border. These highways were said to be fifty feet wide and were built by removing hills and filling seas to make them straight. They were all lined with pine trees. Besides pines, the trees most frequently used for streets and avenue planting in ancient China were willows, sophora, chestnut and elm.

SELECTION AND IMPROVEMENT

In cultivating plants, man has from the very beginning striven to improve on their products. In edible fruits, efforts are made to improve upon their size and flavor. In flowers, improvements are aimed at their size, color, and shape. In ornamental and shade trees, it is aimed at better shape and healthier growth. During this continuous process of improvement through the centuries, new varieties of various species appeared and were preserved and propagated. The cultivated plants are one of our most cherished heritages from the past, the result of unceasing toils of endless generations of farmers and gardeners through the entire history of mankind.

Man has for centuries taken advantage of certain natural laws, without actually being conscious of them, in furthering his efforts to improve his cultivated plants. Nature creates variations in plants by inserting new genetic factors (genes) into the inheritance mechanism, by causing new and abrupt changes (mutations) in these genetic factors, by altering the original setup of this mechanism (chromosomal changes), and by crossing different genetic stocks (hybridization).

Aside from effects of environment, which are not heritable, these genetic changes result in creating differences among plants. No two individuals in nature are exactly alike. The differences vary from the very slight and subtle to the drastic. This is called "variation" in genetics, the science of the study of heredity. Variation is the basis of natural selection, which brings out the very complex phenomenon of organic evolution. Taking advantage of the basic fact of variation, the keen farmer or gardener selects the one plant out of a thousand that possesses certain desirable features and plants and propagates it. This is called "artificial selection," or "plant breeding," which has been carried on through generation after generation and has produced the cultivated plants we have today.

Thus, before modern genetics became a science toward the end of the nineteenth century, improvement on trees had already been made to a certain extent. In most of our common tree species there are a number of varieties and these have been developed mostly in cultivation. Many new hybrids have also been raised and maintained, especially among genera with species distributed naturally in disjunct areas and brought together in cultivation in modern times. Such geographically isolated and genetically related species often readily produce hybrids such as in *Aesculus,* the horse chestnuts, and *Tilia,* the lindens. Hybrids, if fertile and bred among themselves or with their parent species, will not produce constant progeny but will show all intermediate characters ranging from one parent breed to the other. In other words, hybrids will not breed true to type. In tree cultivation, however, the advantage of vegetative propagation can overcome this difficulty. The progeny of a single individual, including those of hybrids, thus propagated, are all uniform in their character, producing a line technically called a "clone," and thus all desired characters of that individual can be perpetuated and multiplied.

In some such hybrids, when effected by certain internal changes, the entire complement of the inheritance mechanism (chromosomes) may become doubled. In this way the plant will subsequently be able to breed true from seed, in fact giving rise to what sometimes is called a "new species." Among the trees an example is *Aesculus* × *carnea,* the red horse chestnut, a hybrid between the common horse chestnut, *A. hippocastanum* of Europe, and the red buckeye, *A. pavia* of North America. Both parents have a chromosome number of forty, but the hybrid has the number of eighty.

The chromosome number in a given species is normally constant. All cells of the plant carry the same number of chromosomes with the exception of the gametophytes and the gametes, the gamete-producing generations and the reproductive cells respectively, in which the number is reduced to one-half. The male and female gametes eventually unite to produce the new embryo and seedling. With regard to chromosomes the number in the embryo and consequently in the mature plant is known as "diploid," meaning two sets, and the number in gametes as "haploid," half sets. When the gametes unite to produce the new embryo, the haploid number is restored to diploid, thus completing the cycle.

Changes in chromosome number occur frequently in nature for various undetectable reasons. The change may involve the addition of one or a few more chromosomes (aneuploidy). It may be multiplications of entire sets of chromosomes (polyploidy) such as the addition of one, two, three, etc., sets to the original two (triploid, tetraploid, pentaploid, etc.). It has been found that such a change in chromosome number can sometimes be induced by artificial means, such as treatment with colchicine.

All such changes in chromosome numbers, like all other changes in the basic genetic pattern (genotype), bring corresponding changes in the appearance of the individuals (phenotype). Whenever new and desirable characters of a heritable nature appear, characters due to genetic and not environmental effects, the keen plant breeder will have a chance to select and perpetuate.

Professor Zirkle (1935) has traced very lucidly the history of plant hybridization in former times in his book *The Beginnings of Plant Hybridization.* He thinks artificial pollination was invented very early, probably antedating the invention of writing, as authentic records of pollination extend back to Babylonian times. Many fruits, such as the date palm and the fig, were hybridized by the ancients in prehistoric times, though only incidentally. But in ancient and medieval times, the sexes of plants were not clearly known and gardeners and argiculturists on the whole were ignorant of the function of pollen. Thus spontaneous hybridization was not early recognized, and was explained as spontaneous degeneration. Accurate accounts of plant hybridization date from the first half of the eighteenth century, but it was apparently practiced earlier. Most of the hybrid trees have been developed, recognized, and selected since that time.

DISPERSION OF CULTIVATED TREES

Plants were known to be deliberately transplanted by men since the earliest historical times. Abraham brought trees from other lands for planting. Solomon collected all kinds of plants and he not only had an orchard of fruit trees trees but also planted on his ground what were called barren trees; among these the cedar was apparently brought from the mountains.

The earliest historic record of foreign plant introduction is in the fifth Egyptian dynasty, dated by modern historians as around 1570 B.C., when Queen Hatshepsut brought foreign seeds and plants from the land of Punt—generally believed to be the east coast of Africa. She sent her ship to this foreign land for seeds and trees so she might produce aromatic incense in her own gardens. She caused sculptors to accompany the expedition, and returned, they carved and set up a bas-relief at Luxor showing gardeners loading the boat with incense trees in tubs and piling its decks with seeds.

One of the seven wonders of the ancient world is the Hanging Gardens of Babylon of the third century B.C. They were planted with exotic flowers, trees, and lianas. Pumps worked day and night to water these celestial gardens, which towered 430 feet into the air. The gardens were built by the King of Assyria as a pleasure seat for his queen Shammuramat, whom the Greeks called Semiramis.

In China, records state that at the death of Confucius in 481 B.C., his disciples, reputed to be 3,000 in number, planted trees in his graveyard as memorials, all brought from their native regions far and wide. This may be called the first arboretum on record, a garden of trees and not just a park or orchard. In ancient China, records of tree introduction from foreign lands by various emperors are quite numerous. Many monarchs built extensive gardens and had trees transplanted from distant lands. In the second century B.C., Emperor Wu Ti of the Han dynasty, for instance, had litchi transplanted, though unsuccessfully, from the newly conquered Annam to the capital in northwestern China.

The story of plant dispersion and introduction has been carried on through all ages and in all countries. Plant introduction has been carried on so naturally and so gradually that few people, even historians, realize the importance of its role in the advancement of human culture. Restless and wandering men have for centuries carried with them seeds and plants, favorite vegetables or fruits or trees that embody a far-off homeland. Plants have been transplanted from point to point on the earth for food, beauty, or mere sentiment.

Relatively few records of plant introduction, however, are recorded in historical chronicles. Numerous instances of introductions, deliberate or accidental, made by merchants, travelers, pilgrims, and adventurers, or brought about by migrations and invasions, never entered into any writing. Our study on plant dispersion has to be made largely by inferences based on circumstantial evidence.

In plant dispersion, the most important route in ancient times was the trade route known as the "Silk Road" connecting the two great centers of civilization in the East and the West through the deserts of central Asia. Peach, orange, and others were brought from China to Persia in very early times. Grape, pomegranate, and others were brought from there to China around the second century B.C. The weeping willow reached Babylonia from China through the same route at an unknown date. After the tenth century, contacts between the East and the West, in addition to this overland route, were also effected by merchant ships sailing across the Indian Ocean.

The discovery of the New World initiated an age of great plant migration. The beginning of this important chapter of human history, however, came with no fanfare. Scarcely two weeks after the arrival of the first settlers on the Island of Jamestown in 1607, the pioneers had carved new habitations out of a virgin land, and were planting seeds of vegetables and trees brought over from the Old World. On the other side of the continent the early Franciscan fathers planted fig, grape, and olive in California, where they still persist today.

Not only were Old World species brought to new grounds for planting, but numerous new species from the hitherto unknown flora of North and Central America were also being introduced into cultivation for the first time. In a brief span of three hundred years, American trees have been planted over the world, and the black locust is becoming the most widely planted tree of mankind.

MODERN BOTANICAL EXPLORATIONS

The great maritime activities from the sixteenth century onward not only resulted in the discovery of many lands in the New World but also in the opening up of vast areas in the Old World formerly forbidden to the outside world. Sailors and adventurers were soon followed by botanical explorers in quest of new plant resources. Their interests were at first largely centered around economic plants and plant products, but later the interest was extended to ornamentals and subjects of pure botanical significance.

The New World, of course, revealed an entirely different flora and vegetation, both as to cultivated as well as spontaneous plants. It is not necessary to emphasize here the great impact on the Old World economy made by such crop plants from the Americas as corn, potato, tobacco, and many others. Numerous trees from America were also introduced and planted, and much altered the scenery of many Old World cities and gardens. Exploration in the New World is still being actively carried on in vast areas in

Central and South America which remain little known botanically.

Botanical exploration in the Old World in the nineteenth and twentieth centuries has also revealed, among other things, the extremely rich flora of temperate eastern Asia in China and Japan. Not only are cultivated plants of economic or ornamental significance numerous there, but the more remote mountainous districts are found to contain flora that is richer than any other part of the temperate world. Many of the species have since been introduced into cultivation and have contributed greatly toward tree culture and horticulture of the entire world.

2

The Weeping Willow and
Lombardy Poplar

The weeping willow and Lombardy poplar are two of the most widely planted trees in the world. In appearance, few trees show such extreme contrast, one being broad, rounded, with drooping branches and the other narrow, pyramidal, with strongly fastigiate branches (Figures 9-13). Yet the two are closely related members of a distinct family, one being in the genus *Salix* and the other in *Populus*, the only two genera of the phylogenetically very much isolated family Salicaceae. Thus, in spite of their extreme outward differences, the two trees are near relatives, sharing between them a common genealogy and heritage.

The origins of cultivation of both trees are, in widely different ways, somewhat shrouded in mystery. The weeping willow, believed by Linnaeus and other earlier authors to be from Babylonia, is now quite definitely known to be a species of Chinese origin, cultivated since ancient times in that country but still growing wild in some parts, and brought from the East in early historical times by human efforts. The Lombardy poplar, on the other hand, is a man-made tree, most probably a selected stock from the common black poplar in Europe, and originated about some three

9. Salix babylonica.

hundred years ago. Its origin forecasts the great emphasis on clonal selection as an important phase of poplar culture now so extensively practiced in both Europe and America.

THE WILLOW FAMILY

The willow family is primarily restricted to the temperate regions of the Northern Hemisphere. It is readily distinguished from all other groups and constitutes the sole

10. *The weeping willow, Salix babylonica.*

11. *The weeping willow in early spring.*

12. Populus simonii 'Fastigiata.'

13. Salix × elegantissima.

family of the order Salicales. The genus *Salix,* the willows, contains some 300 species and *Populus,* the poplars, some 40 species, but owing to frequent hybridization in nature, especially in the willows, their classification into species is exceedingly difficult matter. Their identification is further complicated by the fact that the sexes of the trees are different, and their flowers usually appear before the leaves are developed.

The poplars can readily be distinguished from the willows, in addition to other technical characters, by their drooping male and female catkins, which in the willows are stiff and erect. The development of the catkins is associated with the fact that wind pollination takes place in the poplars, while in the willows pollination is generally by insects.

The Japanese botanist Nakai, who proposed numerous new species, genera, and families in his lifetime, once established a third genus *Chosenia,* referring to it a willow from Korea. This is *S. bracteosa,* which differs from the other willows in its drooping male catkins, its two almost free styles, and its glandless flowers. Rehder (1940) and others have followed him in recognizing this anomalous species as representing a distinct genus. Extensive morphological studies made by Hjelmqvist (1948) and others show, however, that in spite of these more or less distinct characteristics, there exists a continuous series of transitions in other species that renders such a differentiation undesirable.

USES

The willows and poplars have been useful to mankind since prehistoric times. The use of willows for making baskets and containers, probably among the first articles manufactured by man, contributed much to the advance of early culture, and to this day the growing of basket willows is still an important industry in some parts of the Northern Hemisphere. The bark of some species is used for tanning or dyeing. As shade trees, both willows and poplars are invaluable because of their hardiness and rapid growth. As ornamental trees, like many other tree species possessing innate beauty, the most attractive kinds are indispensable for certain kinds of landscape decoration. The trees are also widely planted for windbreaks or screens or as holders of the soil. A few species are utilized locally for timber or for making charcoal.

DISEASES

One of the great advantages of growing willows and poplars is their rapid growth, but these fast-growing trees are also prone to many diseases. Many species are susceptible to canker caused by bacteria, to canker and stem dieback caused by a number of fungi, and to certain leaf diseases caused by rusts. Susceptibility may be increased due to unsuitable growing conditions. Thus under certain unfavorable conditions, some of these diseases may wipe out an entire population.

Among the insect diseases, the most serious one is caused by the poplar or willow borer or weevil, a beetle imported from Europe. This pest is causing extensive injuries, especially in the northeastern United States, to the basket-willow industry and to many poplar and willow trees used for landscape planting.

THE CULTIVATED WILLOWS

Willows have been cultivated in Europe and Asia since time immemorial for baskets, for timber, or for shade. Most of the European species are widespread in their natural ranges, occurring all the way to western, central, or sometimes even to eastern Asia. The cultivation of some apparently dates from prehistorical times and it is impossible to ascertain the exact localities where the different species were first grown.

Among the species cultivated specially for basketry work are the almond-leafed willow, *S. amygdalina,* the common osier, *S. viminalis,* the purple osier, *S. purpurea,* and a form of the white willow, *S. alba* var. *tristis.* The most valuable timber trees of the genus are the crack willow, *S. fragilis,* and white willow, *S. alba.* Other species, like bay willow, *S. pentandra,* goat willow, *S. caprea,* gray willow, *S. cinerea,* and creeping willow, *S. repens,* are cultivated mainly for ornament.

In eastern Asia, the most widely cultivated species are the weeping willow, *S. babylonica* and *S. matsudana,* the latter in more northern regions and in wet as well as dry localities.

Many of the European species were introduced into America in colonial times and quite a number of these have escaped into the wild in the eastern states and have become naturalized. The native American species did not begin to be cultivated until about the beginning of the eighteenth century. Among the better known and early grown species are the pussy willow, *S. discolor,* and silky willow, *S. sericea,* of the northeast, and black willow, *S. nigra,* of wider range. The native basket willow, *S. petiolaris,* is cultivated in the northeast for making baskets along with the introduced osiers, *S. viminalis* and *S. purpurea.*

THE CULTIVATED POPLARS

Many species of poplars have been cultivated in the Old World since very ancient times. Of European and western Asiatic origin are such common species as the white poplar, *Populus alba,* gray poplar, *P. canescens,* European aspen, *P. tremula,* and black poplar, *P. nigra.* Of eastern Asiatic origin are the Chinese white poplar, *P. tomentosa,* and Chinese aspen, *P. adenopoda.* Of African and Asiatic origin is *P. euphratica,* the original "weeping willow" of the ancients.

The association of sadness with poplars seems to have originated early both in the East and the West. It probably alludes to the captivity of the children of Israel "by the rivers of Babylon." Lombardy poplar is often used in graveyards in the place of cypress in some countries of Europe and the Near East. In China, the poplar has been considered since most ancient times as one of the five official memorial trees, the one reserved for the graves of the common people.

The most commonly cultivated American species is the cottonwood, *P. deltoides,* of eastern North America. The tree is very variable and runs into many forms and varieties. In North America it was apparently first cultivated in the first half of the eighteenth century or earlier. It is now the most generally planted tree in the middle states. It was introduced into Europe around 1750, and the first description of it was published in 1755 by Duhamel, who extolled its value for the decoration of parks (Sargent, 1896). In Europe, it is generally planted as a timber tree and is the most widely cultivated species of American poplars. There it hybridizes with the European black poplar, giving rise to numerous hybrid poplars known as *P. × canadensis,* widely used in reforestation work.

Among the more recent introductions from eastern Asia

are some highly ornamental species. Some of these are wild, others have long been cultivated there and were first introduced into the West toward the latter part of the nineteenth century. Among the more notable is *P. maximowiczii,* from northeastern Asia and Japan, a handsome poplar with striking bark and foliage. The bark of old trunks is gray and deeply fissured. The leaves are large, dark green on the upper surface and whitish on the lower. Another handsome species is *P. simonii* from northern China, a tree with slender branches and rather small bright green leaves. There is a variety 'Fastigiata' of narrow pyramidal form with upright branches (Figure 12).

THE WEEPING WILLOW

The weeping willow, *S. babylonica,* has a confused history in the literature. The name "willow of Babylon" was mistakenly bestowed on this tree, which is quite definitely of Chinese origin. The willows mentioned in Psalms growing by the rivers of Babylon are now known to be not willows at all but a poplar, *P. euphratica.*

In China the weeping willow is a favorite garden tree and has been widely planted from time immemorial. It now occurs nearly all over the country, especially along the waters on alluvial areas. As it is extensively planted, it is difficult to tell the wild from the cultivated trees. In western China, in the mountainous region on the upper reaches of the Yangtze River, it is apparently still growing wild, and there trees attaining to the largest size are to be found (Wilson, 1920).

From the most ancient times, the weeping willow has been closely associated with the Chinese people, playing an integral part in their garden art, folklore, painting, and literature. The slender twig swaying freely in the wind is compared in the traditional literature with a dancing figure, hence the common term "willow waist." The weaving branches also symbolize affection and attachment, and willows were at former times picked to give to a departing friend to signify unbroken friendship. The willow tree is an indispensable item in the rock garden, always planted by the side of the pond which usually is the main feature of the design. It is widely known in the West in the "willow pattern" porcelain which is to this day quite popular.

The weeping willow does not grow well in colder countries, especially in drier situations. The willow that takes its place in northern China, around Peking, for instance, is *S. matsudana*. A pendulous form, especially planted for ornamental purposes, it resembles closely the weeping willow in general appearance. Other local species are similarly used in Korea, Formosa, and other areas in eastern Asia.

DISPERSION OF THE WEEPING WILLOW

The weeping willow was introduced from China into Japan in early times and is now commonly planted there in gardens and also in the streets. The tree cultivated is the male form and from there it has also been introduced into California and widely planted around San Francisco (Wilson, 1920).

The spread of the weeping willow toward the West, into central Asia, must have happened at a very early date but nothing is known definitely. Most probably it was carried westward by travelers along the overland "Silk Road" through the deserts, either deliberately or accidentally. As willow branches were used extensively in former times for weaving baskets, crates, and for many other purposes as well, and as the twigs are easily rooted by mere insertion into the ground, incidental introduction and cultivation in many lands could have happened readily and repeatedly.

As mentioned above, the tree introduced into Japan is male while the one introduced into Europe is female. In China both male and female trees are common. The downy seeds, known as "willow downs," flying in the air in late spring are a familiar sight in regions where willows are abundant, and constitute a frequent subject of poetry.

The weeping willow was not known in western Europe until around 1700, at a time when the direct sea route to China was already quite active. However, the first introduction of the tree seems not to have come by sea but indirectly from western Asia, a fact which shows that the species must have been well established in central and western Asia at a much earlier date.

The Englishman James Petiver, in his work *Musei Petrivertiani Centuriae* . . . published in 1703, mentioned that a specimen collected by Cunningham from China has pendulous branches and bears the Chinese name Yan-liu (Bretschneider, 1882-1895). Cunningham went to China by sea in 1698. This shows that the tree must have been unknown to England in cultivation at that time.

A later record of the tree is found in Tournefort's *Corolarium* of 1719, which describes the tree as the Oriental willow, as found in the Levant. The tree must have been finally introduced into England sometime before 1730 for it was known to be on sale in gardens near London as stated in a catalogue published by Philip Miller in that year (Wilson, 1920).

According to Wilson, since the weeping willows in Europe are female they are in all probability originated from a single tree introduced either by Tournefort or Wheeler from western Asia. The latter mentions a tree which might have been a weeping willow in his book *Journey in Greece and Asia Minor* published in 1682, but this record is even more uncertain.

Although Wilson asserts that the tree was probably introduced into England indirectly from western Asia, it seems also highly probable that it could have arrived directly from China by sea in the late fifteenth or early sixteenth centuries. The fact that all European trees are female makes Wilson's contention that they are all decendants of a single tree feasible, but again more than one clone might have been involved.

From Europe the weeping willow was carried to many other parts of the world, including North and South America. It is not known when it was brought to the United States but Wilson thinks it was probably toward the end of the eighteenth century.

The most famous of these cultivated trees is probably one of the many planted on St. Helena by Governor Beatson about 1810, one which became a favorite with Napoleon then in exile, and under which he used to sit at length meditating on his past glories and vanities.

THE WEEPING WILLOW IN CULTIVATION

The weeping willow, like other species of the family, is fast-growing but short-lived. It is widely cultivated for its distinct ornamental effect but aged or very large specimens are not common in cultivation. It is also not hardy in very cold climates.

The willows hybridize freely in nature. The female trees of the weeping willow, as cultivated in the West, have produced in the nineteenth century several hybrids or supposed hybrids with some other species, and these hybrids have since been maintained by vegetative propagation. These trees resemble somewhat the weeping willow in general appearance, especially in the long hanging branches, but are usually more hardy and are thus often planted where the weeping willow proves to be tender.

Two of these trees are supposed hybrids between the weeping willow and the crack willow, *S. fragilis,* of Europe and western Asia. One of them is called *S.* × *elegantissima,* the Thurlow weeping willow (Figure 13), and the other *S.* × *blanda,* the so-called Wisconsin weeping willow. A third tree is a putative hybrid between the weeping willow and *S. alba,* the common white willow of the Old World, and is named *S.* × *sepulcralis.*

THE ORIGIN OF THE LOMBARDY POPLAR

The Lombardy poplar, ever since it was first known, has been a problem to botanists and horticulturists. The tree, as compared with the weeping willow, has a much shorter history, as it apparently first appeared only some three hundred years ago, but its origin is even more mystifying than the weeping willow.

The Lombardy poplar first originated as a cultivated tree in Europe. It was a tree unknown to the ancients. It was not recorded by Pliny, who certainly would have noted it had it existed then.

The earliest cultivated trees seem to come from northern Italy, where the black poplar is common, particulary in Lombardy, on the banks of the Po, whence the name Lombardy poplar or Po poplar. The Italians generally called the tree "cypress poplar," *pioppo cypresso,* on account of it shape.

Manetti noted in 1836, as cited by Loudon (1838), that

along the banks of the Po, there were Lombardy poplars growing among the black poplar, *P. nigra*. As these appeared on exposed banks which had been flooded, he believed the seeds were buried in the soil for many years and came from extinct forests. This seems hardly likely and, as noted by Loudon, it would be more plausible to consider these as seeds of the current year carried there by the wind. Manetti's letter, apparently a translation from the Italian, is the only reliable record of this tree made on the basis of first-hand observation at this early date. It was reprinted in the *Gardeners' Chronicle* in 1883, but as the anonymous author notes therein, "it is very contradictory and vague in some parts."

The Lombardy poplar is considered by some earlier authors as native to Persia and the Himalayan region and probably carried from there to Italy (Loudon, 1838; anonymous, 1883), but there does not seem to be evidence to substantiate these views. Because of the fact that it was not introduced into Tuscany till 1805, Loudon believes that it is not indigenous to Lombardy or any part of Italy, as he thinks it certainly could not escape the notice of the Roman agricultural writers.

All evidence seems to indicate that the Lombardy poplar first made its appearance in the Po Valley of Italy. The enigma which confronted Loudon can be explained if we assume that it did not exist in ancient times, but appeared, some three hundred years ago, as a mutant of the black poplar. Its remarkable shape accounts for its selection and cultivation by some enterprising farmers and its subsequent dispersion to other countries.

That the Lombardy poplar is closely related to the black poplar was a fact very early noted by botanists. It was first considered by Munchhausen as *P. nigra* var. *italica* 1770; later as a distinct species under such names as *P. italica*

Moench 1785, *P. dilatata* Aiton 1789, *P. pyramidata* Moench 1785, and *P. fastigiata* Desfontaines 1804. However, the closeness of the two in vegatative as well as reproductive structures leaves little doubt that the Lombardy poplar is, in spite of different growth form, only a variation of the common black poplar of Europe.

This seems to be the generally accepted view at the present. In a recent monograph on poplars by Hesmer (1951), it is called *P. nigra* var. *italica*. As this tree represents a form selected for cultivation, the modern terminology renders its as *P. nigra* cv. 'Italica' and it is so named in another recent monograph issued by the FAO (1956).

THE DISPERSION OF THE LOMBARDY POPLAR

From Italy, the Lombardy poplar was introduced into France in 1749 and at about the same time also into Germany (Loudon, 1838). It was introduced into England at about 1758 from Turin (Aiton, 1789). Since then it has been generally planted throughout Europe, either as a roadside or ornamental tree, or occasionally as a timber tree. The date of its introduction into America is given as 1784 by Sargent (1896). During the nineteenth century, the tree was taken to other parts of the world from Europe or America. It is now very commonly cultivated in eastern Asia, as a street or shade tree or for reforestation purposes.

THE LOMBARDY POPLAR IN CULTIVATION

The Lombardy poplar is now one of the most widely planted trees around the world, but nearly all planted trees are male (Figures 14, 15). Only a few female trees are

14. *Lombardy poplars in New York State. (Photo, courtesy of Ernst Schreiner.)*

15. *Lombardy poplars along highway in Ontario, Canada. (Photo, Canadian Government Travel Bureau.)*

known to exist, and these have less strictly upright branches and consequently a broader head. The occurrence of both male and female trees shows a number of clones are involved.

In cultivation, there is another fastigiate form of the black poplar, cv. 'Plantierensis,' very similar in habit to the Lombardy poplar but usually not so slender. This resembles closely the Lombardy poplar and is generally also called by that name. It differs, however, in the hairy branches and leaf stalks, and for this reason, it is supposed to be a cross between cv. 'Betulifolia,' downy black poplar, and the true Lombardy poplar. This tree originated at Plantières, France, in about 1885 or earlier, and both male and female plants are known.

The main value of the Lombardy poplar is ornamental. The distinct fastigiate shape is often exploited with admirable effect in landscaping to contrast with horizontal lines of architecture and with round-headed trees. It often fits in the landscape to balance horizontal lines likes bridges or walls, or certain types of buildings. Its taller, pointed heads give contrast to the round heads of most other trees.

The Lombardy poplar, like other species of poplars, is subject to many diseases. It is generally resistant to bacterial canker and seems to be fairly resistant to rust. However, the trees, especially the older ones, suffer rather violent dieback due to some as yet unknown cause. This dieback is fairly common in European trees and was especially prevalent some eighty years ago, but it is now of more frequent occurrence in North America.

This disease condition was formerly believed by some as a decline in vigor due to clonal senescence (Focke, 1883). However, the European trees are now in an apparently better condition than before the turn of the century, and thus evidently this is not the case. More probably, in the

Lombardy poplar, a number of clones were actually represented, and as the more susceptible ones had been killed and eliminated, the surviving ones are among the more resistant.

The prevalence of this destructive disease makes this tree unpopular in North America. Attempts to introduce from Europe certain resistant clones should be made in order to revive the Lombardy poplar populations in America.

BREEDING AND SELECTION OF POPLARS

In recent years, great efforts have been made in Europe as well as in the United States in the promotion of research and culture of poplars. The improvement of poplars is aimed especially at the production of fast-growing, disease-resistant strains. The technique includes primarily hybridization and clonal selection. (Pauley, 1949; Schreiner, 1959).

The native species of poplars have been used for centuries in all parts of Europe. Hybrid poplar culture was first started there about 200 years ago when the American cottonwood was introduced. Countless numbers of hybrids have since been produced by natural hybridization between the introduced American species and the native European black poplar, followed by backcrosses between the hybrids and the parent species.

Many of these hybrids or putative hybrids are superior to either one of the parental species in rate of growth, disease resistance, stem form, and other desirable characteristics. As these poplars can be readily propagated by cuttings, superior individual trees can be selected and perpetuated.

The most important hybrid poplars are crosses between the American cottonwood, *P. deltoides,* and the European black poplar, *P. nigra,* known collectively as *P.* × *canaden-*

sis. These hybrids were probably originated first in France before 1750, and there are now many different forms, named and unnamed, propagated as clones. These were originated independently in different places between the various forms of the two parent species.

The culture of poplars is now continued on a more scientific basis. Instead of selecting from spontaneous hybrids, controlled breeding is now performed and clones are selected and propagated from the resulted hybrids bearing desirable characteristics.

3

The Plane Trees

The familar plane tree, the most widely planted street and shade tree in many American and European cities, has been the subject of much confusion. The three commonly planted species, the Oriental plane, the American plane, and the London plane, because of their generally similar and at the same time extremely variable foliage characters, are often wrongly identified, not only in trade catalogues but also in horticultural literature. The confusion is further aggravated by the uncertain origin of the London plane, which has baffled horticulturists and botanists for many decades.

The London plane is unknown in the wild state and apparently originated in cultivation (Figure 16). In horticultural literature before the early years of the present century, it was generally considered as a variety of the Oriental plane, and because of this, when the Oriental plane, or *Platanus orientalis,* is mentioned in the literature, it may be either to the true Oriental plane or to the London plane, *P.* × *acerifolia.* The prevalent consensus is that the latter is most probably a hybrid between the Oriental plane and the American plane, *P. occidentalis,* but the confusion does not abate (Figure 17).

Some of the confusion is probably due also to the many common colloquial names applied to these trees; these

16. Platanus × *acerifolia.*

names will be discussed below. For the present discussion, the common names used for the three species in question will be Oriental plane for *P. orientalis,* American plane for *P. occidentalis,* and London plane for *P.* × *acerifolia* (Figures 18-20).

The Oriental plane is one of the oldest and most cherished shade trees cultivated by men, and is still widely planted in many parts of the world. The American plane came into cultivation during the sixteenth century, while the London plane originated as a cultivated tree about

17. Leaves of Platanus occidentalis (left), P. × acerifolia (middle), and P. orientalis (right.)

18. Platanus orientalis.

19. *Platanus occidentalis.*

20. *Platanus × acerifolia.*

the later part of the seventeenth century. These three plants are among the most prominent features of the landscapes in many habitats of man's choice, from cities of modern skyscrapers in America and Europe to towns made of mud huts among the steppes of western and central Asia. The story of the plane trees indeed is most instructive in illustrating vividly the influences exerted on each other by man and the surrounding vegetation.

To trace the history of cultivation of the different plane trees, especially the mysterious and elusive London plane, it is necessary first to take into account the genus *Platanus*, which, aside from the fact that it contains some of the most attractive and valuable shade trees, is a genus that will repay consideration.

THE GENUS PLATANUS

The genus *Platanus* is the only member of the family Platanaceae, a family of special interest because of its isolated position and disputed relationships in the phylogeny of the flowering plants. The eight or nine species of *Platanus* are scattered in widely disjunct areas in the Northern Temperate Zone, a pattern of distribution which generally indicates antiquity. The family history of *Platanus*, is truly majestic, for paleobotanists trace its ancestry back to the Age of Dinosaurs in the late Cretaceous, some 100,000,000 years ago.

The species of the genus *Platanus* more or less closely resemble each other in general appearance. They are tall, deciduous trees with smooth bark shredding in broad, brittle, thin plates and exposing the whitish or brownish inner bark. The leaves are large, palmately lobed with toothed or smooth margins and with the leaf stalks enlarged at the base enclosing the winter buds. Stipules usually sheath the base of the leaf stalks, and are of varying sizes and often fall off very early. The flowers are of distinct sexes produced on separate inflorescences on the same tree. They are very minute and simple, without perianth and closely packed in hanging globose heads. The fruiting heads remain attached to the tree during the winter on the otherwise bare branches, presenting a graceful appearance. These heads break up in late winter and the hairy fruits are widely distributed by winds. The wood of these trees, light brownish or reddish in color, splits poorly and is thus not highly valued.

There is one species in southeastern Europe and western Asia, *P. orientalis,* the Oriental plane. In eastern North America, the American plane, *P. occidentalis,* occupies a wide range from southern Maine to Ontario and Nebraska south to northern Florida and Texas. The variety *glabrata* occurs from central Iowa and Missouri to western Texas and northern Mexico.

In western North America *P. racemosa* occurs in southern California and Lower California, and *P. wrightii* from New Mexico and Arizona to California and northern Mexico, sometimes planted in the latter region as a shade tree. Also known from Mexico are *P. mexicana* (also occasionally planted as a shade tree), and *P. lindeniana.* In addition, Standley proposes two species from Mexico, *P. chiapensis* and *P. oaxacana.* Thus seven species are recorded as occurring in Mexico, but some of them are local or only little known (Standley, 1922). Inasmuch as the species are generally differentiated by leaf shapes, and as these are extremely variable even on the same tree, it seems likely that as a result of critical monographic studies, fewer species will be recognized.

Palebotanists have found that in the Tertiary many

species of the genus were widespread through the northern part of the Northern Hemisphere, in all of Europe, northern Asia and North America north to the Arctic Circle (Seward, 1931). Subsequently the Pleistocene glaciation exterminated the more northern populations, and surviving species are now confined to the eastern Mediterranean region in the Old World and to eastern and western North America, including Mexico. Inasmuch as a large number of such Tertiary survivals are now found existing mostly in eastern North America and in China and Japan, the entire absence of *Platanus* from eastern Asia seemed to be very remarkable (Berry, 1923).

However, in 1939, the French botanist Gagnepain (1939) made an interesting discovery: a plant was found in Laos, Indo-China, which had all the peculiar features of *Platanus* in the flowering structures but with unlobed, entire-margined, and pinnate-veined ovate-lanceolate leaves. In spite of the striking difference in the vegetative structures, Gagnepain considered the plant as belonging properly to the genus *Platanus* and named it *P. kerri* (Figure 21).

The Indo-Chinese species has a longer inflorescence than all the other species, with nine to eleven heads in the fruiting clusters. In the other species, the fruiting heads vary from one in the American *P. occidentalis,* to two to six, in the Oriental *P. orientalis* and others. The larger number is considered a more primitive characteristic and in this sense the Indo-Chinese species probably represents the most primitive species of the genus in existence.

If this proves to be the case, as detailed studies of various other aspects of the plant will verify, the unlobed, entire-margined leaf may indicate a more basic type than the palmately lobed ones, long considered as distinctly characteristic of the genus. This discovery may thus prove

21. Platanus kerri.

of profound significance in the study of the fossil history
of the flowering plants, as the genus *Platanus* has long
been recognized as one of the most important and abundant of the earlier fossils of the flowering plants. Since
fossil plants are usually identified by the shape and venation of detached leaves only, such identifications, as is well
known among paleobotanists themselves, are frequently

not quite reliable. The leaves of this Indo-Chinese species, if discovered in rock strata as fossil imprints, will never be recognized as belonging or even related to *Platanus*. This species, which apparently has not yet been noted by many paleobotanists, indicates that a revision in our interpretation of fossils pertaining to *Platanus* is to be expected.

THE ORIENTAL PLANE

As noted above, the Oriental plane occurs in southeastern Europe and western Asia, eastward to Kashmir, but the range has apparently been extended through long years of cultivation. Henry (1908) believes that it occurs wild in Albania, Greece, Cyprus, Crete, Rhodes, and Asia Minor, while the occurrence in the wild state elsewhere, such as in Iran, Afghanistan, Kashmir, etc., is very doubtful. He considers the wild form to differ slightly from the cultivated form in the slightly smaller leaves with cuneate instead of truncate or cordate leaf base, but as the range of variation in the shape of the leaves in both the wild and cultivated forms is considerable, it is not easy to distinguish the two morphologically. The trees cultivated in England or elsewhere are derived from trees indigenous to Greece and Asia Minor. The leaves of the trees cultivated in Kashmir and Iran are much larger, with broad oblong-triangular segments, and, according to Henry, perhaps represent a distinct race.

Since very ancient times, the Oriental plane has been valued as an ornamental shade tree. Its shelter-giving qualities, the wide-spreading branches and large dense foliage, render it one of the most prized trees in the hot lands of the Near East. With its massive size and great age, it is one of the noblest of all trees and many romantic legends are attached to it.

The plane tree was known in the earliest records of Greece. Herodotus tells that Xerxes, when he invaded Greece, was so enchanted with a beautiful plane tree that he encircled it with a collar of gold and confided the charge of it to one of his Ten Thousand. Aelianus adds that Xerxes passed an entire day under its shade, compelling his whole army to encamp nearby, and that this delay was one of the causes of his defeat. He was so fond of this tree that he covered it with gold and gems, styled it "his mistress, his minion, his goddess," and for several days was entirely oblivious to his expedition and army (Loudon, 1844).

In the time of Pliny, plane trees were planted near all the public schools in Athens. Pliny says that there is "no tree whatsoever which so well defends us from the heat of the sun in summer." We are told that the groves of Epicurus, in which Aristotle taught his roving disciples, the groves of Academus, in which Plato delivered his celebrated discourses, and the shady walks planted near the Gymnasia and other public buildings of Athens, were all composed of this tree. Homer frequently mentions "the shady plane." Socrates swore by the plane tree, and this was one of the things that offended Melitus, who thought it a great crime to swear by so beautiful a tree (Loudon, 1844).

The plane tree was also cultivated in Iran from the earliest period, and is still one of the most conspicuous features of its landscape. The Romans were long attracted to this precious and beautiful tree of the Levant. We are told how Licinius Mucianus, when Roman Consul in

Lycia, dined in its hollow trunk along with eighteen persons of his retinue. It was introduced into Italy from Greece about 390 B.C., and the Romans planted it extensively in their gardens for shade.

The Oriental plane seems to have been introduced into England as an ornamental tree about the middle of the sixteenth century, the exact date being unknown. It was introduced into France from England in about 1754 and into North America apparently in colonial times, although the exact date cannot be ascertained.

However, at the present, aside from the Mediterranean region, the Oriental plane is little planted because of its tenderness. It is now uncommon in Great Britain, where it was replaced by the London plane more than 150 years ago. However, there are scattered fine old specimens of large size of this species planted some 300 years ago in England (Elwes & Henry, 1908).

THE ORIENTAL PLANE IN NORTH AMERICA

There seems to be no record extant fixing the date of introduction of the Oriental plane into North America. Although the Oriental plane sometimes figures deceptively as an important tree of frequent occurrence in this country, actually it is now extremely rare in cultivation. What is generally called *"Platanus orientalis,"* or "Oriental plane," in the trade and in the literature nearly always proves to be the London plane, *P.* × *acerifolia.* The Oriental plane is indeed a plant about which there has been much confusion in horticulture.

This confusion seems to stem from the early concept of considering the London plane as a variety of the Oriental plane. Many authors in the past called the London plane simply *"P. orientalis,"* without indicating, or knowing, that it actually was *"P. orientalis* var. *acerifolia."* Although Henry pointed out, as early as 1908, that the London plane, widely planted in the United States, was invariably known by the erroneous name of *"P. orientalis,"* the confusion in the names persists to this day in the literature. In 1916 Rehder (in Bailey, 1916) noted that "The true oriental plane is rare in cult., the tree usually planted under this name being *P. acerifolia."* Henry also stated in 1919 that the Oriental plane, which is not readily propagated by cuttings, was never used for planting in streets in Europe or North America, and that it was vary rare in the latter region.

In a recent work, Wyman (1951)[1] recommends both *P. orientalis* and *P.* × *acerifolia* as trees for American gardens and states that both are planted annually as a street tree by the thousands and that both of these species have been grown as clipped screens and arbors in this country. Despite inquiries from various sources, we have not been able to confirm these uses of *P. orientalis* in this country.

As a result of these inquiries and from a study of herbarium materials and the literature, we have come to note only a few alleged trees of *P. orientalis* in this country. In New York City, Croizat (1937) located four trees, two standing in Central Park, one planted in Morningside Park, and one cultivated in the Brooklyn Botanic Garden. He says that all of these are mature plants probably of the same age and same origin, although he gives no indication of either. An accompanying illustration and several pre-

---

[1] In Wyman's book there is mention of trees of *P. orientalis* in New England said to be three or four hundred years old. Upon inquiry, Wyman admits this to be an error, stating that he meant England instead of New England.

served herbarium specimens show that these trees are probably correctly identified, although they may also be hybrids resembling more closely *P. orientalis*. He says that the leaves of the Oriental plane vary bewilderingly: Actually some of the leaves of his Oriental plane are very close to those of his London plane given in the same illustration.

In the city of Philadelphia, no less than 158,000 trees are planted along the sidewalks. According to Mr. W. B. Satterthwaite, principal aboriculturist of the city, 50,000 are plane trees, and all except two are London planes. The two exceptions are probably Oriental plane, growing along Verree Road in the Fox Chase section of the city (Figure 22). Another tree of its kind growing nearby as a street tree is in the town of Ardmore, Pennsylvania.

A resident along Verree Road informs us that the plane trees there were planted some twenty-five years ago from seedlings obtained from the nearby Krewson's Nursery. The two supposedly Oriental plane trees are somewhat distinct from the numerous other London planes, being much more slender and more sparsely branched. The outer barks are nearly completely peeled off exposing the smooth, silvery-grayish inner bark. There are three to five fruiting heads to the cluster. The leaves are deeply five- to seven-lobed, with the sinuses reaching below the middle of the leaves. As the trees are trimmed every year we are not aware of any of the younger branches having been killed by frost.

In addition to the three trees on the records of Fairmount Park we have discovered a fourth tree, on Paper Mill Road just beyond the city limits near Chestnut Hill. It has the same kind of deeply and narrowly lobed leaves and smooth, whitish bark. It is a larger tree than those on Verree Road, with the trunk approaching the size of the

22. Platanus × acerifolia, with deeply lobed leaves; from planted tree, Verree Road, Philadelphia.

other London plane trees lining the street along with it. These trees are about the same size and age as those on Verree Road and were probably all planted about the same time. The tree along Paper Mill Road, however, bears fruits mostly in clusters of twos.

These four Philadelphia trees, like the trees of New York mentioned above are, in all probability, not pure *P. orientalis* but hybrids of *P. orientalis* and *P. occidentalis*. However, instead of showing intermediate characters between the two parent species like most other London planes, they resemble more closely the former parent. Our supposition is prompted in part by the fact that although the leaves of these trees approach very closely *P. orientalis,* the smooth, whitish bark seems to resemble more *P. occidentalis* than it does the former. Without definite records of the origin of these trees, it is not possible to ascertain their true identity, and since they approach so closely the features of *P. orientalis,* they may pass as that species for practical purposes in identification even though they may be of hybrid origin. These plants appear to be especially close to *P. orientalis* var. *cuneata,* which is considered by Henry (1919) as a variety of *P. × acerifolia.* (See further discussions under London plane.)

In recent years there have been several efforts made in this country to introduce authentic stocks of *P. orientalis* from Europe and the Orient. The Morris Arboretum received in 1954 seed from three sources, from Kashmir through Mrs. Laura Barnes, from Turkey through Dr. Frank Meyer, and from Italy through Dr. Benjamin Blackburn. A number of seedlings were raised but all perished either in the greenhouse or when outplanted, except one in the nursery which is now about two feet high. This tree is from seed obtained from Italy (Figure 18).

THE AMERICAN PLANE

In eastern North America, *P. occidentalis* occurs in a very wide range from Maine to Ontario westward to Minnesota and southward to Florida and Texas. Within the region it is a common tree, inhabiting especially the borders of streams and lakes and rich bottom lands. A variety *glabrata,* with smaller, more deeply lobed leaves, but considered as indistinct by some botanists, occurs in central Iowa and Missouri to western Texas and northern Mexico.

ORIGIN OF THE LONDON PLANE

The London plane apparently first appeared in cultivation in the seventeenth century. Various opinions were expressed regarding its origin and there were attempts to trace it to some Asiatic sources. While it was first generally regarded as a variety of the Oriental plane, from the very beginning suggestions of its hybrid nature appeared from time to time in the literature.

Henry made a thorough study of this problem (1908). At first he rejected the hybrid origin of *P.* × *acerifolia,* but his subsequent investigations (1919) led him to believe firmly in the hybrid origin of the London plane. Such is the view now generally accepted. Henry is of the opinion that the remarkable vigor of the London plane, its power of resistance to drought, smoke, and other unfavorable conditions of soil and atmosphere is due to it hybrid origin.

Henry believes that the London plane must have originated as a chance seedling in some botanic garden, where an American plane and an Oriental plane happened to grow close together. His historical research led him to surmise it was possibly originated in the Oxford Botanic Garden at about 1670.

Henry's surmise was based on the following: The American plane was introduced from America to England by Tradescant in 1636, while the Oriental plane arrived about a century earlier. By 1670, the American species would be old enough to bear pollen. There is a manuscript left by Jacob Bobart, Jr., curator of the Botanic Garden at Oxford in 1680, which is without date, although a similar one bears the date 1666. In the enumeration of planes in cultivation, besides *P. orientalis* and *P. occidentalis,* a plant is listed as "P. inter orientalem et occidentalem media," and corresponding to this description of a "Plane intermediate between the Oriental and American species," there is a herbarium specimen, undoubtedly *P. × acerifolia,* in the Sherard Herbarium at Oxford, labeled "Platanus media." Specimens labeled "Platanus media, n.d. Bobart, Ox." are in the British Museum, made at about the same period, together with the type specimen of the first published description of the London plane by Plukenet in 1700.

Thus, it is the opinion of Henry that the original tree, which Plukenet describes as bearing large fruit balls in 1700, was then living in the Oxford Botanic Garden, and as it may have been then thirty years old, this would give the date of origin of *P. × acerifolia* as 1670.

Henry's surmise, which as he says is something which cannot be definitely proved, may not be far from the truth. It gives, however, only the earliest historic records of the tree, and does not necessarily mean that subsequent trees of the London plane are all derived from this single individual, as implied by Henry.

Henry, moreover, points out earlier (1908) that the *P. occidentalis* of *Evelyn's Sylva,* ed. Hunter, 1683, is undoubtedly *P. × acerifolia.* This shows that there might possibly be more than a single source of origin even at this early date.

A point against Henry's assertion that *P. × acerifolia* originated in England is the fact that *P. occidentalis* does not live long enough to flower in England and that there is no record that the tree ever flowered there in former times (Bean, 1919). Although the earliest record of *P. × aceri-*

folia is from English sources, it is nevertheless quite possible that actually it might have originated concurrently or exclusively on the European continent, where both *P. occidentalis* and *P. orientalis* are hardy.

The first record of the London plane on the continent was made by Tournefort in 1703 as "P. orientalis aceris folio." Since then the cultivation of this tree has spread all over the continent.

An early opinion on the origin of this plant was expressed by Miller (1731), who says of *P. × acerifolia,* "Although by some supposed to be a distinct species from either *P. orientalis* or *P. occidentalis,* it is no more than a seminal variety of the first; for I have had many plants that come up from seeds of the first sort which ripened in the Physick Garden (Chelsea), which do most of them degenerate to this third sort; which in the manner of its leaves, seems to be different from either, and might reasonably be supposed a distinct kind by those who have not traced its original."

The current scientific name of the plant is adopted from the one given by Aiton in 1789, who considers it as a variety of the Oriental plane as *P. orientalis acerifolia* (in *Hort. Kew.* 3:304, 1789). Willdenow is the first to accord it specific status as *P. acerifolia* (Ait.) Willdenow (in *Linn. Sp. Pl.* ed. 4, 1:474, 1805), but, in spite of this, most authors up to the beginning of the current century treated it as a variety of *P. orientalis.*

The London plane is a tall tree with an upright stem, attaining a size intermediate between the taller *P. occidentalis* and the slightly smaller *P. orientalis.* The bark usually exfoliates in large flakes as in *P. orientalis* and the inner bark is variable in color. The leaves are large, with five broad triangular lobes closely resembling those of *P. occi-*

dentalis but in general with the middle lobe slightly narrower. The lobe is about as long as it is broad, while in *P. occidentalis* the breadth of the middle lobe exceeds its length. The fruiting heads are extremely variable in size and number on the peduncle: mostly in twos, but varying to as many as six. The individual achenes are similar in structure to those of *P. orientalis* and do not resemble those of *P. occidentalis.*

It can thus be seen that *P.* × *acerifolia* resembles *P. occidentalis* more closely in foliage and *P. orientalis* in fruiting structures. Because of the close similarity of the leaves between *P. occidentalis* and *P. orientalis,* the two were much confused with each other in the literature of the eighteenth and nineteenth centuries. Inasmuch as *P. occidentalis* is quite unsuitable to the climate of England and Europe, it must have been very rare in these countries in former times (as at the present) and plants referred to as *P. occidentalis* were invariably *P.* × *acerifolia.* Not until 1856 did Hooker (1856) clarify the confusion by distinguishing between the two on the basis of fruiting characters, which had hitherto been unnoticed.

The hybrid nature of the plant, although not definitely recognized at first, was long suspected by many authors. A horticultural name, *P. intermedia* Hort., of unknown origin but expressing the apparent hybrid nature of the plant, long appeared in the synonymy of *P.* × *acerifolia.* It is listed under *P. orientalis* var. *acerifolia* Ait. by Loudon in 1844,[2] and might have existed much earlier.

---

[2] This antedates Rehder's assertion (Rehder, 1949) that the name, in synonymy, is to be credited to the *Kew Handlist of Trees and Shrubs* of 1896. Still less understandable is the listing of this name by *Index Kewensis* Suppl. 9, 1938, attributing this name, also in synonymy, to Chow, *Familiar Trees of Hopei*, 1934. This latter work follows essentially the nomenclature as given in the first edition of Rehder's *Manual* of 1927.

That the hybridization did not occur in one instance only is indicated by the independent recognition of this phenomenon in several other countries. Before 1731, there was known in England a plant called the Spanish plane tree which is essentially similar to *P. × acerifolia* but with leaves having more cordate bases. The description, in Miller's *Dictionary,* 7th ed. (1759), is based on a tree planted in 1731 as a variety accidentally arisen from seed. Miller's tree is unmistakably the Spanish plane, later known as *P. hispanica.* This rare tree, which originated some time before 1731, was, in the opinion of Henry, probably a seedling of one of the early London planes. The same tree was imported to England from France in 1856 under the name *P. macrophylla* (Rivers, 1860).

In 1804, Brotero (*Fl. Lusitan.* 2: 487. 1804), describes *P. hybridus* from plants cultivated in Portugal. Although this is generally but doubtfully referred to the synonymy of *P. occidentalis,* there is a strong indication that it should be considered as a synonym of *P. × acerifolia,* which has a year's priority over the latter name.

These various records seem to indicate that hybrids between *P. occidentalis* and *P. orientalis* appear to have arisen independently in several different countries at different times. It may be argued that the Portuguese and Spanish plants were introduced from England, but inasmuch as they are quite distinct from the English types, this seems rather unlikely. On the contrary, the Spanish plant, which was called Spanish plane in England as early as 1759 or earlier, must have come from Spain rather than the reverse. Furthermore, *P. occidentalis* is known to be tender in England, seldom reaching the adult fruiting stage there, but is more at home in southern Europe. In the latter area hybrids between it and *P. orientalis,* which is also more commonly planted there than in England, are more likely to occur than in any region further north.

The Spanish plane, also known as *P. macrophylla* as illustrated and described by Bean (1919) from trees grown in England, shows more distinctly the influence of *P. occidentalis* in its fruiting characters than other plants of *P.* × *acerifolia.* The fruiting heads are occasionally two or three together, but usually solitary, with the individual achenes almost smooth and more flattened at the apex than those of *P. orientalis.*

VARIATIONS OF THE LONDON PLANE

From the records, it seems therefore justifiable to conclude that *P.* × *acerifolia,* or the London plane, is an assorted group of trees of variable characters which have originated as hybrids between *P. orientalis* and *P. occidentalis* independently in different countries in western Europe and the British Isles at various times. Since the two species hybridize, it is very unlikely that hybridization happened only once and not in other localities, especially where more individual trees of both species are present. The two parent species concerned are very variable in their characters. This explains the occurrence of the many different types of hybrid progenies derived from those two species. These progenies, whether propagated vegetatively or as seedlings, maintain or further increase the variation of the complex known collectively as *P.* × *acerifolia.*

Among the cultivated specimens of *P. orientalis* in Europe, there are usually recognized two varieties, *cuneata* and *digitata.* Although their leaves resemble more closely *P. orientalis,* Henry is apparently right, in view of their frequently imperfect achenes, in considering them as of hybrid origin. Henry believes them to be second-generation seedlings of *P.* × *acerifolia.* Although we have no way of ascertaining their origins, they are possibly cultivated clones of advanced generation segregates of the hybrid *P. × acerifolia* that resemble most closely *P. orientalis* in leaf shape.

Henry (1919) recognizes *P. × acerifolia* as the first-generation hybrid of *P. orientalis* and *P. occidentalis* and accepts six additional "species" as second-generation hybrids, including *cuneata* and *digitata,* mentioned above, and two new species of his own as follows: *P. hispanica* Muenchhausen, *P. pryamidalis* Rivers, *P. cuneata* Willdenow, *P. digitata* Gordon, *P. cantabrigiensis* Henry, and *P. parviloba* Henry.

According to modern taxonomic concepts and nomenclatural practice, since *P. × acerifolia* is considered as a hybrid between *P. orientalis* and *P. occidentalis,* it should be designated either by the formula *P. orientalis × occidentalis* or by the name *P. × acerifolia.* Only one specific epithet is admissible in this case, and such other specific names as given by Henry, if considered as of hybrid origin between the two same parental species, should all be included in this concept. Variations within *P. × acerifolia* should be recognized as cultivars (cv.) and rendered in roman in quotes instead of italics. Such cultivars as are propagated solely by vegetative means from a single known stock are known as clone (cl.). The recognized variations of *P. × acerifolia* (Henry & Flood, 1919; Rehder, 1940) are described and discussed below.

(1) *P. × acerifolia* cv. 'Hispanica'

Leaves large, often 30 cm. in width, persistently tomentose on the nerves and petioles, shallowly cordate or cuneate at the case, with five distinct short, broadly triangular lobes, the margins dentate. Heads not so flattened as in *P. occidentalis,* and not so conical as in *P. orientalis.*

This taxon appeared formerly also under the name *P.*

× *acerifolia* var. *hispanica, P. hispanica,* or *P. macrophylla.* It is cultivated in England and France, but is apparently unknown in America. It is a vigorous tree, which, according to Henry, produces good seeds from which seedlings, which are not uniform, can be easily raised; however, in nurseries it is invariably propagated by cuttings.

(2) *P.* × *acerifolia* cv. 'Cuneata'

A small tree with leaves that are deeply five-lobed, very cuneate at the base, conspicuously dentate at the margins, practically glabrous when adult. Two to four fruiting heads, small (rarely over 2 cm. in diameter), composed of few, often imperfect achenes.

This is a variety long placed in *P. orientalis,* as young trees of cultivated *P. orientalis* and even certain adult wild forms often bear cuneate leaves and are scarcely distinguishable. But Henry points out that this variation is distinguished by its fruiting heads, which are small and often made up of few, mostly imperfect achenes and for this reason he believes it to be probably of hybrid origin. It conceivably originated sometime before 1789, when it was first made known by Aiton.

(3) *P.* × *acerifolia* cl. 'Digitata'

Leaves resembling *P. orientalis* but much smaller, to 13 cm. broad, the base truncate but with a short central cuneate part, with five lobes, elongated, dentate, with wide sinuses. Two or three fruiting heads, very small (about 1.2 cm. in diameter), composed of a few imperfect achenes.

Henry is probably right in sugg!esting this as originated from a seedling of *P.* × *acerifolia,* and not a variety of *P. orientalis,* as designated by most authors. According to him, it is rare in cultivation, and only two trees are known, both in England.

(4) *P.* × *acerifolia* cl. 'Pyramidalis'

Of upright habit, the lower branches widespread but not drooping when old. Leaves truncate, usually three-lobed, the lobes broadly triangular, slightly dentate.

This clone was probably originated in France about 1850. It is now planted in England as a street tree.

(5) *P.* × *acerifolia* cl. 'Kelseyana'

Leaves spotted in the center with yellow, the margins green. Known also as *P.* × *acerifolia* var. *aureo-variegata*.

(6) *P.* × *acerifolia* cl. 'Suttneri'

Leaves large, white over most of the surface but with green spots in the center.

(7) *P.* × *acerifolia* cl. 'Cantabrigiensis'

Leaves rather small, to 13 cm. wide, the base truncate with a cuneate central part, five-lobed, the lobes distinct, short, triangular, entire or with one or two teeth. Three fruiting heads, small (about 2 cm. across), of few imperfect achenes.

A single tree of unknown origin in the Cambridge Botanic Garden, described by Henry as *P. cantabrigiensis*.

(8) *P.* × *acerifolia* cl. 'Parviloba'

Leaves variable, truncate to cuneate, with three triangular or oblong-triangular slightly dentate or entire lobes and two additional small lobes or teeth near the base. Three to six fruiting heads, small (about 2 cm. across), made up of a few achenes, some imperfect.

A single grafted tree at Kew of unknown origin; described by Henry as *P. parviloba*.

THE LONDON PLANE IN AMERICA

The London plane, as well as the Oriental plane, must have been introduced into America from Europe during colonial times, although records of its early introduction do not seem to exist. As mentioned above, the London plane and the Oriental plane were often not distinguished in the horticultural literature of the nineteenth century and earlier, and what was called the Oriental plane was in all cases the London plane. The confusion between the two seems to persist up to the present, especially in the trade, while in Europe the London plane was formerly much confused with the American plane.

In Philadelphia, there is a generally circulated notion that William Harper, presumably raising them from seed, first introduced the London plane into this country about fifty years ago. A published record of this is found in R. H. True's report of Jackson's lecture on the diseases of the plane tree in 1936 (True, 1936): "It was reported by Mr. S. N. Baxter, City Arboriculturist of Philadelphia, in the discussion following the lecture, that this type of plane tree (London plane) was introduced into Philadelphia about thirty years ago by Mr. William Warner Harper, of Andorra Nurseries." Mr. Harper is apparently the first one who introduced this tree on a large scale, but earlier than this the London plane must already have been in existence in this country, especially around Philadelphia, as there are references to it, for instance, in *Meehan's Monthly* in the late nineteenth century. The tree was then usually known as the "Oriental plane." Incidentally, at Andorra Nurseries, the London plane is today still called "Oriental plane" and since they propagate it solely by cuttings, their present nursery stock may be the direct descendants of the early trees introduced from London some fifty years ago.

In North America, the Oriental plane is not hardy in the North, but the London plane is hardy as far north as southern New Hampshire. As noted above, the true Oriental plane occurs rarely if at all in cultivation anywhere in North America, and trees bearing this name are nearly all London planes. Actually nearly all the plane trees extensively planted along the streets throughout the country are London plane, as the native American plane, because of its susceptibility to fungus disease, is not satisfactory for this purpose.

The London plane, as cultivated in this country, appears to be extremely variable. In and around Philadelphia, observations made on numerous trees in parks and along streets show that marked variations are readily discernible in the color of bark, shape of leaves, and number and size of fruiting heads. The inner bark varies in color from brownish to silvery-grayish and the outer bark peels off in patches of different sizes and in different degrees, although more often than not the patches are relatively large. The leaves are exceedingly variable in size and shape, including the base, the number, size, and shape of lobes, the serrations, and the hairiness. The fruiting heads are mostly in twos, but the number varies from one to three and occasionally more.

The general shape of the leaves of the London plane in America, however, seems to differ from those in Europe, especially in older trees. The lobes of the American trees are broad and short, and while they are slightly longer than those of the American plane, *P. occidentalis,* they are much shorter and broader than those of *P. orientalis.* The sinuses are broad and shallow (Figure 20). In the older trees in Europe (for instance, the London plane tree near the Rue Jussieu entrance of the Jardin des Plantes, Paris, and a tree at Kew), the leaves are deeply and narrowly lobed, approaching very closely the condition in *P. orientalis.* The Paris tree bears the label "Platanus × acerifolia, orientalis × occidentalis, connu depuis 1670. Un des 3 pieds plantés par Buffon entre 1784 et 1788." These trees

bear leaves resembling very closely the few supposed specimens of *P. orientalis* of Philadelphia and New York and
for this reason it is suggested that these American trees
may also be segregates of hybrids between *P. orientalis* and
P. occidentalis. Like the other London planes they happen
to be extremes more closely resembling one of the two
parents and are not true or pure *P. orientalis.*

It is, however, entirely clear that the London planes
now planted in America are significantly different from
the London plane trees in Europe, especially those planted
one or two hundred years ago. The trees in Europe show
strong affiliation with *P. orientalis,* particularly in the shape
of the leaves. The plants generally grown in America show
more closely and strongly characters of *P. occidentalis.*
While a more definite solution to this problem requires
detailed analysis of specimens in quantities, our observation so far points out that most of the London planes in
America are probably not first-generation hybrids of *P.
orientalis* and *P. occidentalis,* but are raised from hybrids
of these species, backcrossed, perhaps successively for
more than one generation, to the native American *P. occidentalis.*

The importance of London plane to modern cities can
hardly be overemphasized. As mentioned before, out of
158,000 trees planted along the sidewalks of Philadelphia,
nearly one-third are London plane. In New York City,
2,282,000 trees are under the care of the Park Department, a large proportion being London planes, especially
in Manhattan, where the London plane, numbering
47,000, forms the bulk of the tree population.[3]

The London plane originated and spread under man's
tutelage less than three hundred years ago. Its dramatic

[3] *New York Times,* August 26, 1955.

use by man parallels the rapid development and expansion of modern cities, including New York and Philadelphia, within the same age. The mutual dependence of the London plane and modern city dwellers constitutes an important and interesting chapter in anthropological chronicles. Many modern urbanites are not aware of the fact that as their forefathers removed the natural vegetation of the land to create cities, it was largely the London plane that took the place of forest trees. These trees of the city streets not only give shade and beautify their surroundings but also are reminders of the changing seasons of the year and purify the very air they breathe.

COMMON NAMES OF PLANE TREES

Confusion in the status of cultivated plane trees can be attributed in part to the numerous common names applied or misapplied to these plants, which, in turn, reflect their variable nature and puzzling origins.

The name "plane" derives from the scientific name *Platanus*, adopted from the classical name derived from the Greek *platys*, meaning "ample," in allusion, according to most authors, to its spreading branches and shady foliage. Fernald (1950), however, says that *platys* apparently refers to the large leaves. It is known as *Platanus* in German and *platane* in French. In America, the common name of the different plane trees is sycamore, but in Europe this name is always applied to *Acer pseudo-platanus* and never to the plane. In Scotland the name plane tree is also applied to *Acer pseudo-platanus*. In reality the sycomorus of the ancients is the *Ficus sycomorus* of northeastern Africa.

Oriental plane is the name generally applied to *P. orien-*

talis in England and America, although in the literature it is occasionally also known as eastern plane. It is called *Morgenlandischer Platanus* in German and *platane de l'Orient* in French and is known to the Persians as *chinar* and to the Arabs as *doolb* (Loudon, 1844).

The American plane, *P. occidentalis,* is also known in the horticultural literature of England as the Occidental, or western, plane. In America, besides being called sycamore, it is also commonly known as "buttonwood," and sometimes by such less common names as "cotton tree," "water beech," or "buttonball tree.":

The London plane and its variations have acquired a long list of names in less than 300 years since their first appearance. In the literature, the following names, most of them no longer in use, are recorded: park plane, Spanish plane, London plane, sycamore, maple-leaved plane, European plane, Oriental plane, and eastern plane. The last two names were applied to the London plane in earlier literature, when it was generally considered as a variety of *P. orientalis,* and this indiscriminate use, as mentioned before, is a main source of the confusion of the two plane trees in subsequent years, in both the trade and in horticultural literature, down to the present.

4

The Ginkgo

No plant more than the ginkgo has a stronger claim to be called to "living fossil," a term used by Darwin to designate survivals of the past. This very same tree was thriving a hundred and twenty-five million years ago when dinosaurs were still roaming the earth, and the genus has remained from that time until now almost unchanged. Its ancestry traces back for another hundred million years to the later part of the Permian period in the late Paleozoic. These ancestral and related plants were most abundant during the Upper Triassic and the Rhaetic-Jurassic ages in the early Mesozoic. Thus the sole living member of a once great and dominant race of the vegetation of the world, the ginkgo is, among all the tens of thousands of plant species existing today, a most precious and tenuous link between the present and the remote past.

From its last natural refuge in the mountains of eastern China, the ginkgo has now spread through cultivation to many parts of the world. With the aid of man, it now reinhabits grounds lost to nature during the past million years. It thrives even in a modern urban environment which many other trees cannot tolerate. It defies all pests. A tree that survives the vicissitudes of a hundred million years must have come from a most tenacious stock and have been specially selected by mother nature out of thousands of similar and dissimilar plants of the same age.

The ginkgo, with its spreading, rigidly ramified branches and curiously shaped leaves, is a tree of great distinction and dignity in appearance. Human history appears insignificant when compared with the genealogy of this tree which now stands in the gardens and along city streets created by man who appeared millions of years later. As an eminent paleobotanist, the late Sir Albert Seward (1938), in discussing its geological history, remarks: "It appeals to the historic soul: we see it as an emblem of changelessness, a heritage from worlds too remote for our human intelligence to grasp, a tree which has in its keeping the secrets of the immeasurable past."

The ginkgo survives as a living plant in China only. Although there is a great mass of writing on this tree in the Western languages, it does not seem that the history of its origin and cultivation in the Orient has ever been properly investigated. There are the often-repeated but somewhat misinformed statements that the ginkgo has been in cultivation in China since very ancient times, that its origin is entirely unknown, and that it is preserved in cultivation only by Buddhist monks on monastery grounds.

This chapter reviews briefly the horticultural and botanical history of the plant, its historical and geographical origins, early cultivation in China, subsequent introduction into Japan and the Western world, botanical investigations made on the plant, and problems pertaining to its scientific name.

ORIGIN IN CHINA

The ginkgo was unknown to the ancient Chinese; its records in the literature cannot be traced definitely beyond one thousand years. The earliest Chinese civilization prospered along the Yellow River Valley in northern China.

Many plants have been domesticated there since time immemorial; these as well as numerous wild plants are recorded in the earliest classics and other writings of three thousand years or more ago. Among the earliest domesticated trees are the peach and apricot; the wild ancestors of the former have long since disappeared. Numerous other trees like pine, arbovitae, juniper, jujube, etc., are mentioned in this ancient literature, trees that are still native to northern China. But the ginkgo is conspicuously absent, attesting to the fact that it was not a native tree of that region in ancient times, although at present it occasionally occurs there as a cultivated species.

Before the Sung dynasty, in the late tenth century, the ginkgo was almost unknown in northern China. It first appears in literature in the eleventh century as a plant native to eastern China, south of the Yangtze River. During the Sung dynasty, the Yangtze River Valley became more and more developed and more closely integrated with the north. Because of pressure exerted by the invading Chin Tartars from the north, there was a gradual southern movement of culture and population. This movement was particularly intensified around 1127 A.D., when the capital was moved from Kaifeng in northern central China to Hangchow in eastern China south of the Yangtze River. Before this, the period is known as the Northern Sung (960-1127 A.D.) and from then on until the end of the dynasty, when the Mongols conquered the whole of China and established the Yuan dynasty, as the Southern Sung (1128-1279).

EARLIEST RECORDS

In the eleventh century, the ginkgo, when it first appears

in the literature, was known as a plant of the region south of the Yangtze River in eastern China. At Kaifeng, the capital, it was considered as a rare and precious "fruit." It was sent as tribute yearly from its native region to the capital to be presented to the emperor. Subsequently a few trees were planted in the capital, and this is the first recorded instance of cultivation outside of its natural range. Gradually it became more commonly grown and its cultivation also spread to other regions.

Previous to the Sung dynasty literature, there are only a few alleged but unproven references to the ginkgo. In a poetical essay, entitled *Wu Tu Fu,* on the capital of the Wu Kingdom, by Tso Ssu of the Tsin dynasty (265-420), there is mention of a fruit called *p'ing chung kuo.* The Wu Kingdom is modern Kiangsu and Anhwei provinces, the native habitat of ginkgo. A later commentator noted that the fruit is of a silvery color. For this reason, herbalists and botanists of subsequent ages, such as Li Shih-chen in 1596, and Wu Chi-chun in 1848, suggested that this might be the ginkgo but none of these authors regards it as certain.

The ginkgo is also mentioned in a work called *Chung Shu Shu, (The Book on the Culture of Trees),* a small volume on agriculture attributed to the famous gardner Kuo To Tu, or Kuo the Hunchback, of the eighth century. However, this is actually a work of much later origin, probably of the Yuan dynasty around the fourteenth century. That the ginkgo is mentioned in this work as *yin hsing* (silver apricot), a name apparently not in use earlier than the late eleventh century, indicates its much later origin.

In another instance the ginkgo is possibly referred to in a story attributed to a T'ang poet in a Sung notebook. Thus the ginkgo might have been known to the northern

Chinese before the Sung dynasty, but it was apparently a very rare plant, not then in cultivation.

CULTIVATION IN THE SUNG DYNASTY

There are many definite records about this plant in the Sung literature. It was, as mentioned before, first sent as tribute from its native place, mentioned specifically as Suancheng of the Ningkuo District in what is now the southern part of Anhwei province, south of Wuhu, a port on the Yangtze River. In a Sung notebook entitled *Shi Hua Chung Kuei,* it is said that "In the Capital (Kaifeng), there was originally no *ya chio* (ginkgo). Since Prince Li Wen-ho came from the south and transplanted it to his residence, it becomes famous. From then it gradually propagates and multiplies, and fruits from the south are no longer considered precious." Prince Li lived in the first half of the eleventh century.

Another Sung work, entitled *Chun Chu Chi Wen,* mentions that "During the reign of Yuan Feng (1078-1084) there was a fruit in the capital called *ya chio tzu* (ginkgo). There were four big trees producing several bushels of fruits each year."

The name *ya chio,* meaning "duck's foot," refers to the shape of the leaves. This is evidently the earliest name of the plant, and possibly the name first known in its place of origin. After the nut became known in the capital Kaifeng, another name, *yin hsing,* or "silver apricot," referring to the "fruit," was adopted, and the two names were often used concurrently.

GINKGO IN SUNG POETRY AND PAINTINGS

The Sung dynasty was a period of great artistic and

literary activity. During this period, the ginkgo was not only first introduced into cultivation, but it also began to appear in poetry and in paintings. It was eulogized by many famous poets, who mostly praised its "fruits" and sometimes also its leaves. Undoubtedly the praise of the rare "fruit" by the many renowned poets brought fame to this plant in northern China.

Among the verses we may mention especially the ones by the famous poets Ou-yang Hsiu and Mei Yao-ch'en because of their historic interest. Ou-yang (1007-1072), a famed historian, essayist, and poet, was a native of Kiangsi province. Mei (1002-1060), also a famous poet, was a native of southern Anhwei province. Both then held official positions in the capital Kaifeng.

Their exchange of poems began when Ou-yang presented Mei with some ginkgo nuts from the trees of Prince Li mentioned above. This nut, then rare and considered precious in the capital, was familiar to Mei as it was produced in Mei's home region. In expressing his gratitude to Ou-yang he wrote a verse which says that *ya chio* (duck's foot) was unfamiliar to the northerners just like the walnut to the southerners, and that it was so named because of its leaves. It was produced in Mei's native region, Suancheng. He was thus happy to find it planted in the capital as a rare and precious "fruit" and was sentimental in receiving this as a present since it originated in his home area.

Ou-yang was a great historian, and on this occasion he took the opportunity in answering Mei with a poem in which he wished to leave some useful information for posterity. The contents of the verse can be translated simply as:

"*Ya chio* (duck's foot) grows in Kiangnan, with a name which is not appropriate. At first it came in silk bags as a tribute, and as *yin hsing* (silver apricot) it became cherished in the middle provinces. The curiosity and effort of the noble Prince (Li) brought roots from afar to bear fruit in the capital. When the trees first fruited they bore only three or four nuts. These were presented to the throne in a golden bowl. The nobility and high ministry did not recognize them and the emperor bestowed a hundred ounces of gold. Now, after a few years the trees bear more fruits. The friendly owner presents me with these nuts like giving me pearls. In the past Chang Chien (second century B.C.) introduced grape and pomegranate (from central Asia). We can imagine that when these first came, they must have been similarly highly valued as these nuts. But now these plants are common all over China, growing along fences and walls. The very things are still the same, but human nature changes in time. Someone should record the beginning so that future generations can know its origin. This is thus not only continuing your verse, but also contributing to history."

Subsequently they further exchanged poems when Mei returned to his home town and sent nuts to Ou-yang. In these poems Ou-yang mentioned that Mei's nuts were collected from the forests in the wild and both poets now used the name *yin hsing,* or silver apricot, but not *ya chio,* or duck's foot.

The ginkgo was the subject of many verses by other poets during the Sung dynasty. It also became a frequent subject for painting. In the *Hsuan Ho Hua P'u* (*Catalogue of Paintings of the Imperial Collection*), of the reign of Emperor Huei Tsung (reigned 1101-1125) there are listed several paintings of this plant. One of them, entitled "A picture of *yin hsing* (silver apricot) and the bird *pei tou wen,*" was by an artist Yo Shih-hsuan, an official of the reign of Emperor Shen Tsung (1067-1084). Another entry lists two paintings of "A study of *ya chio* (duck's foot)" by Prince Tuan Hsien Wang, fourth son of the Emperor Yin Tsung (reigned 1063-1066) and brother of Emperor Shen Tsung.

It is interesting to note that the earlier interest in the plant is because of the edible nut and the ornamental features of the plant. The "fruit," of course, is not a true fruit, but a drupelike seed. It is yellowish when mature, of the size and appearance of a small apricot but with a silvery bloom on the outside, hence the literary name *yin hsing,* or silver apricot. The inside kernel is edible. This nut is frequently compared by Sung poets with the walnut, a popular nut fruit of the north.

THE GINKGO IN LATER DYNASTIES

There are many herbals and materia medicas in the T'ang and Sung dynasties, but none of them mentions the ginkgo. Ginkgo first appeared in the herbals of the Yuan dynasty (1280-1368) in *Shih Wu Pen Ts'ao,* or *Edible Herbal,* by Li Tung-wan, and *Jih Jung Pen Ts'ao,* or *Herbal for Daily Usage,* by Wu Jui. In Wu's book it is noted that the nuts, if eaten in excess, especially by children, may be slightly poisonous.

Besides the names mentioned, in the Yuan dynasty there were other names in common usage, such as *pei kuo* (white fruit), *pei yen* (white eye), *ling yen* (spirited eye), *jen hsing* (nut apricot), etc. Among these, *pei kuo,* or white fruit, is still the most widely used colloquial name throughout the country in modern times, while *yin hsing,* or silver apricot, remains as the common literary name.

Another common name that appeared in later times is *kung sun shu,* or grandfather-grandson tree. This attests to the fact that only old trees bear fruit and therefore a tree planted by a man will be useful to his grandson.

The ginkgo is a dioecious tree, that is, the male and female flowers are borne on separate individuals and hence only the female tree bears fruits. This phenomenon was considered mysterious by the earlier observers. They early noted, however, that there are male and female trees and the two should be planted nearby to ensure fruiting, but they also entertained the strange idea that female trees if planted alone by the side of water can bear fruit also.

After the Sung and Yuan dynasties, the ginkgo seems to have been widely cultivated all over the country. The Sung and Yuan authors do not mention specifically that the tree occurred only in cultivation including its native habitat, nor is there any association of this tree with religious institutions. The writings of these authors seem to imply that the tree occurred indigenously in southern Anhwei. And it has always been the custom of the Buddhist as well as the Taoist priests to preserve portions of the natural forest around their temples, thus preserving many native species which otherwise would have perished in agricultural areas. Thus, venerable specimens are often found on temple grounds, but they occur in many other places as well. Very old trees, ginkgos as well as many other kinds, are often revered and preserved because of geomantic beliefs, and they are sometimes, especially in the remote interior, decorated with incense stands or wayside shrines. It is a kind of primitive nature worship, and the tree is honored because of its age and not its kind. Contrary to the much-perpetuated belief, emphatically stated especially by Wilson (1920) and frequently copied by other horticultural writers, the ginkgo has no association with Buddhism. The cultivation of the ginkgo tree has no religious origin or significance and its preservation is not attributable to planting by Buddhist or Tòaist priests.

The ginkgo is described in detail in the great herbal *Pen Ts'ao Kang Mu* of 1596 by Li Shih-chen and also in the great botanical work *Chih Wu Ming Shih Tu Kao* of 1848 by Wu Chi-chun, where a good illustration of the foliage and fruit is found.

Li repeatedly says that the ginkgo originates from Kiangnan, a province of the Ming dynasty embracing the modern provinces of Kiangsu and Anhwei. On the name of the plant he says that it was originally called *ya chio* (duck's foot) because of the shape of the leaves and that when it was first sent as tribute during the early Sung dynasty to the throne, the name was then changed into *yin hsing,* or silver apricot, because of the whitish "fruit." His description of the plant, which is detailed and quite accurate reads:

"*Yin hsing* (silver apricot) grows in Kiangnan, especially around Suancheng. The tree is twenty to thirty feet high with thin and straight-veined leaves like the web of a duck. The leaf has a notch at top and is green above and palish beneath. The flowers are borne in clusters in the second month and are rarely seen by people. A single branch often bears tens or hundreds of seeds resembling the seeds of *lien* (*Melia*). The seeds become ripe after frost, and the flesh is removed for the nut inside. The nut is pointed at both ends. Those with three ridges are male and two female. The kernel is green when young, later becoming yellow. Male and female trees should be planted nearby to ensure fruiting, or the female trees can be planted along water. Another method to ensure fruiting is to graft a male branch onto the trunk of a female . . ."

The statement that the number of ridges on the nut can be used to distinguish sexes appears in several other and earlier works. Nuts with three ridges are rare. There does not appear to be much truth in this observation.

Ginkgo is now a widely grown ornamental and nut tree in most parts of China. Occasionally it is made into a dwarf tree for pot culture. The kernel is eaten roasted, or sometimes cooked with fowl or meat. There are also certain medicinal effects attributed to the kernel.

INTRODUCTION INTO JAPAN

The ginkgo was introduced from China into Japan at a fairly early date, perhaps in the Sung dynasty when the tree first became known in northern China. In Japan there are specimens alleged to be nearly a thousand years old. The tree is known there as *icho* and the fruit *ginnan,* the former the corruption of the Sung Chinese named *ya chio* (duck's foot) and the latter the Chinese *yin hsing* (silver apricot). (Koidzumi, 1936; Makino, 1951).

The exact date of introduction from China to Japan is unknown. Wilson's statement (1920) that "the tree reached Japan with Buddhism in the sixth century of the Christian era" is entirely unfounded, as the ginkgo was at that time not even yet known in China. And, as stated elsewhere, the tree has no association with Buddhism whatsoever except incidental planting on temple grounds. It seems to have been carried from southern Kiangsu and Anhwei, the original habitat of the tree, directly to Japan. The Japanese pronunciation of the names of the trees is closer to the Wu dialect of southern Kaingsu than to the northern Mandarin.

Moreover, there is in Japan a legend about growing a ginkgo tree from a cutting in southern Kiangsu in China. According to the *Gazette of Kunshan,* a district in southern Kiangsu in China, this is the story of a ginkgo tree in the Sung dynasty. It is said that when Emperor Kao Tsung moved from Kaifeng in the north to Hangchow in the south in 1127, the imperial cavalcade crossed the Yangtze River into southern Kiangsu. Coming to a town called Chen I near the city of Kunshan between Soochow and Shanghai, an official named Kung I, a native of the northern capital Kaifeng, picked a branch of ginkgo, sticking it into the ground and prayed that if the branch lived, he would settle there. The branch later developed into a huge tree and in later years the trunk became gnarled and twisted and adorned with many hanging "nipples" as in other venerable trees of this same kind.

In both China and Japan, there are now many specimens of immense size. On very old trees there often develop peculiar burrs which are called nipples, *chi-chi* in Japanese. These are due to abnormal development of dormant or adventitious buds. If these hanging burrs reach the ground, they take root and bear leaves (Fujii, 1895). Henry (Elwes & Henry, 1906) gives a picture of such an old tree in western China which is reproduced here (Figure 23). This tree grows on the grounds of a Taoist and not a Buddhist temple.

INTRODUCTION INTO THE WEST

The ginkgo was first made known to the western world by Kaempfer, a surgeon in the employ of the Dutch East India Company, who first observed it in Japan in 1690 and published in 1712 a description with an illustration of the foliage and fruit (Kaempfer, 1712). The tree was first introduced into the Botanic Garden at Utrecht. Jacquin brought it into the Botanic Garden at Vienna sometime after 1768. It was introduced into England about 1754 (Henry, 1906).

23. *A ginkgo tree in western China. (From Elwes & Henry, Trees of Great Britain & Ireland, 1906.)*

Most of the earlier trees raised in Europe appear to have been males. The first recorded female tree was one found by De Candolle near Geneva in 1814. Scions from that tree were grafted upon a male tree in the Botanic Garden at Montpellier, where the first perfect seed is reputed to have been produced.

The ginkgo was first introduced into America in Philadelphia. As far as authentic records go the oldest tree in this country is the one in Woodlands Cemetery, in West Philadelphia, which was brought by William Hamilton

24. *The ginkgo tree in Bartram's Garden, Philadelphia. (Photo, W. R. Carpenter.)*

25. *Ginkgo tree in Tyler Arboretum, near Philadelphia. (Photo, W. R. Carpenter.)*

from England in 1784. This is a magnificent specimen of a male tree. Harshberger (Wilson, 1920) is of the opinion that the ginkgo tree in the old Bartram Garden in West Philadelphia is the oldest and the first planted in America because the garden is older than that founded by Hamilton and the tree is larger (Figure 24). There are other large specimens elsewhere in or around Philadelphia. An old tree in the Tyler Arboretum was planted in 1849 (Figure 25). Another old tree in Germantown, a male, has a female branch grafted onto it, leading many to believe that the ginkgo is monoecious. In the Morris Arboretum, the largest tree is a female, which was planted over seventy years ago, and measuring 7.5 feet in girth. This tree has fruited for a number of years and now bears a heavy crop of seeds every season.

The ginkgo is now probably the most esteemed street tree in this country, especially valued because of its upright habit and freedom from insect pests. It is considerably planted in Philadelphia, Washington, D. C., New York, and many other large cities, especially in the east. It is ironical that the oldest living species of trees is better suited to the most modern manmade habitat than almost any other tree.

BOTANICAL INVESTIGATIONS

The name ginkgo was given by Kaempfer. Kaempfer's designation was adopted by Linnaeus, who in 1771, described the plant as *Ginkgo biloba*. In 1797, the English botanist J. E. Smith (1797), considering the name "uncouth and barbarous," renamed it *Salisburia adiantifolia*. However, the latter name has no nomenclatural standing, as personal reference is not a valid reason to reject an older legitimate name.

When the ginkgo was first known to science, it was regarded as one of the conifers and generally included in the Taxaceae. In 1895, the Japanese botanist Hirase (1895-1898) made the startling discovery that the ovules are fertilized by motile sperm cells conveyed to them by pollen tubes. This differs radically from all other conifers, taxads, and flowering plants, which have nonmotile male nuclei. Motile sperm cells are found only in the lower plants and in the ferns and cycads. This discovery established definitely the unique nature of the ginkgo, which was raised to ordinal rank by Engler.

Hirase's find is often claimed as one of the great discoveries in botany in the nineteenth century. His discovery, followed by further investigation by other botanists, stimulated many detailed studies on the various interesting features of this most unusual plant.

The ginkgo is a deciduous tree. The branches are beset with short spurlike shoots which bear clusters of fanshaped leaves and flowers. The venation in the leaf is open and dichotomous. The male and female flowers are produced on different trees. The male flowers appear in pendulous catkins of numerous loosely arranged stamens. The female flowers arise usually in pairs, each a long stalk with two naked ovules. The seed is drupelike, with a fleshy outer covering enclosing a woody shell with a kernel (Figure 26).

All these structures and many other interesting details have been intensively studied by many botanists, and especially the unusual process of fertilization noted above and the nature of the ovule. The ovule bears a collarlike rim at the base, a structure that has given rise to much discussion (Fujii, 1896; Shaw, 1908). The development of the seed is also very unusual. The seed is normally a mature ovule containing a fully developed embryo. Here

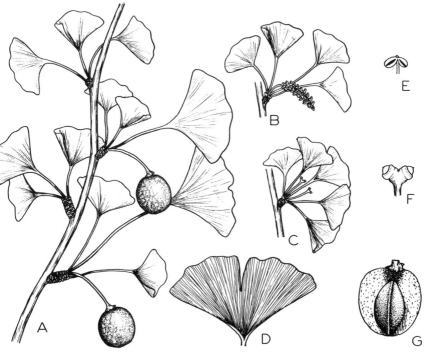

26. *Ginkgo biloba. A. A fruiting branch, B. Branch with male catkin, C. Branch with female flowers, D. A leaf showing venation, E. A stamen, F. A female flower, G. Section of seed showing nut.*

in the ginkgo the ovule drops before fertilization and the embryo is formed later in the detached ovule as it lies on the ground during the winter (Eames, 1955). The equally remarkable development of the embryo was first described by Strasburger (1872). The botanical aspects of *G. biloba* have been monographed by Seward and Gowan (1900) and Sprecher (1907).

The female tree, because of the disagreeable odor of the fleshy covering of the fallen seeds, is less desirable than the male tree in cultivation, especially as a street tree. It is claimed that there are some differences in the growth habit of the two sexes, the male trees being more pyramidal and the female more spreading, and that the male trees tend to loose their leaves earlier in the autumn. These characters, however, do not appear to be constant enough to be relied upon. As it takes many years for the trees to flower, it would be highly desirable to devise some way of telling the sexes in young trees. Recent studies have shown that there is some difference between the chromosome apparatuses of the male and female trees. If a simple technique can be developed to study this chromosome phenomenon, it will be of great help in eliminating young female trees in horticultural plantings.

The study of the only living survival of a very ancient and once dominant group of plants aids greatly our understanding of the past history of vegetation on the earth. Without this living material many paleobotanical problems would remain unelucidated. The fossil history of the ginkgo and its relatives, outside the scope of this review, is a subject much discussed in paleobotanical works. For a detailed and authoritative treatment, the work of Seward (1919) may be consulted.

GEOGRAPHICAL ORIGIN

Since the middle of the last century, when China was opened to the exploration of the botanical collectors, most of these explorers were anxious to discover the native haunts of this famous plant, but though there were claims of finding this tree in the wild state, none seems to have received any recognition.

Among the earlier and also the widest claims was the one

made by the noted traveler Mrs. Bishop toward the end of the last century. In various publications, as quoted by Masters (1902), by Henry (1906), referring to a letter of hers to the *Standard*, August 17, 1899, and by Seward (1911), referring to her book *Untrodden Paths in Japan,* she reported that she had observed it growing wild in Japan in the great forests in the northern island of Yezo, and also in the country at the sources of the Great Gold and Min Rivers in western China. Her claim, however, was seriously questioned by Henry, Seward, and other authors. Wilson (Rehder & Wilson, 1913; Wilson, 1916) is of the opinion that the tree in the forests of Japan and western China that Mrs. Bishop took for the ginkgo was *Cercidiphyllum.* Wilson states that he had never met with a spontaneous specimen, and in his opinion the tree no longer exists in the wild state.

All botanists and foresters of Japan deny that the ginkgo is indigenous in any part of Japan, but rather that it was definitely introduced from China. Explorers sometimes observed the plant in parts of China in natural woods but these were often passed as escapees from cultivation or only in a semiwild state. Often to deny the assertion that it still occurs wild in China, the statement to that effect made by Wilson, because of his extensive experience in China, is usually quoted.

However, Wilson's exploration in China, though extensive, covered only a small part of the country. He collected especially in the Lushan area in northern Kiangsi in eastern China and more widely in western Hupeh and Szechuan. As most of the curious conifers in China and Japan, of which there are many, have a very limited range of distribution, it is unsafe to make a categorical statement about their range and occurrence. For instance, the recently discovered and much discussed *Metasequoia* occurs in a small area in western China within the region extensively searched by Wilson on his many trips but narrowly missed by his itenerary.

Henry pointedly notes in 1906 (Elwes & Henry, 1906) that "Its native habitat has yet to be discovered; and I would suggest the provinces of Hunan, Chekiang, and Anhwei in China as likely to contain it in their yet unexplored mountain forests." None of the many modern plant explorers working in China during the last hundred years was familiar with the historical records of the ginkgo to guide them in their explorations, and Henry's statement, attesting his wide knowledge of the Chinese flora and his sagacious observations, is closest to the truth. As noted above, early Chinese records clearly indicate that southern Anhwei, especially the region of Ningkuo and Suancheng, is the area where the ginkgo originated. Both are now names of cities, but formerly Ningkuo was also the name of a larger district with modern Suancheng as its leading city. In modern plant exploration, the most reliable reports of the ginkgo as a wild plant also have come from this area, the mountain region along the border of Anhwei and Chekiang.

Frank Meyer, botanical explorer for the United States Department of Agriculture, reports [1] that "the ginkgo grows spontaneously near Changhua Hsien, about 70 miles west of Hangchou in the Chekiang province, China." There "the trees are so common that they are cut for firewood." Wilson (1916) says that "it is, however, by no means certain that this is the original home of the ginkgo as these trees may all have descended from a planted tree." But he says further that "Meyer's discovery, however, is interesting, for there is no other evidence of the ginkgo growing spon-

---

[1] The quotations are from Wilson (1916), who does not cite the source. I have not been able to trace these to any of Meyer's publications.

taneously or that it is cut for any purpose."

The area visited by Meyer lies in the western part of the Tien Mu Shan Range, along the northwestern border of Chekiang and southeastern Anhwei. It is in immediate continuation with the Ningkuo and Sauncheng area where early Chinese records place the original home of the ginkgo. The highest mountain of this range reaches 5,075 feet.

Wilson's assertion that these ten square miles of ginkgos may have all descended from a planted tree seems to be highly improbable. Moreover, this area was visited subsequently by other Chinese botanists, and they also noted the same extensive occurrence of the ginkgo tree in a spontaneous state. In the twenties and thirties, ginkgo was observed and collected in that area repeatedly by Cheng, Tsoong, and others (Cheng & Chien, 1933), and Cheng notes especially that "This tree is very common in Tienmushan, growing in association with coniferous and broad leaved trees. It seems to grow spontaneously in that region."

In this area and the surrounding regions, the ginkgo also commonly occurs as a cultivated tree, more frequently so than in other parts of China. This is borne out by my own observations, as I lived in southern Kiangsu for many years and also traveled extensively in northern Chekiang and southern Anhwei, although not in the Tien Mu Shan area. As early as a hundred years ago, Fortune (1847, 1857) observed enormous specimens near Huchow and Ningpo in northern Chekiang and also in southern Kiangsu. In the nearby Chuki District, Tseng (1935) records a number of varieties of ginkgo trees cultivated for their nuts.

Southeastern China, especially southern Anhwei and Chekiang, is the home of many rare plants. Among the conifers and taxads, this is the native ground of the monotypic endemic *Pseudolarix* and *Nothotaxus,* and also *Torreya grandis, Carya cathayensis,* and other rare species. It is the refuge of many relict plants. The case of the ginkgo is very similar to *T. grandis* and *C. cathayensis,* which occur also in this area both wild and cultivated (for their edible nuts). Undoubtedly the last refuge of the ginkgo lies also in this area, a fact not only sustained by historic records but also corroborated by the observations of modern botanical explorers (Figures 27, 28).

The ginkgo is sometimes also observed in a supposedly wild or semiwild state in other parts of China, especially in the mountains of the southwest, in Kweichow, Szechuan, etc. It may be that the natural range of the ginkgo is fairly wide and scattered, but undoubtedly the area in southern Anhwei and northern Chekiang in eastern China is the center of origin of all the cultivated trees, both in China and Japan as well as in other parts of the world.

THE NAME GINKGO

In recent times there have been considerable discussion and dispute on the name ginkgo. Moule (1937, 1944), Thommin (1949), and others believe that as it was apparently intended by Kaempfer as a romanized version of the Chinese ideographs *yin hsing,* it should be more correctly transliterated into the Japanese *"Ginkyo."* Some botanists, such as Pulle (1948) and Widder (1948), in following this view, consider *Ginkgo* as an "unintentional orthographic error" and replace it with *Ginkyo.* On the other hand Henry (Elwes & Henry, 1906) and Barclay (1944) suggested that the name might be a transliteration of the Chinese ideographs *yin kuo* (silver fruit).

The prevalent Japanese name of the tree is *icho* and the nut *ginnan.* Moule, who examined Kaempfer's essential manuscripts in the British Museum and found nothing like

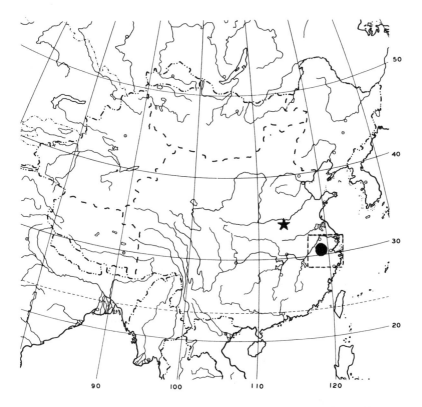

27. *Map of China. Black dot = original habitat of the ginkgo; star = city of Kaifeng; dotted square = area covered by Figure 28.*

the word ginkgo in these, is of the opinion that "according to Kaempfer himself, to the books, and modern evidence, the name of the tree is icho and of the fruit ginnan; while the scarcely known form ginkgo (ginkyo) seems to have got into *Am. Exot.,* and thence into scientific botany, by some slip or misprint which has not been explained."

The case, it seems to me, can be readily settled by a study of Kaempfer's original text, which some authors, such as

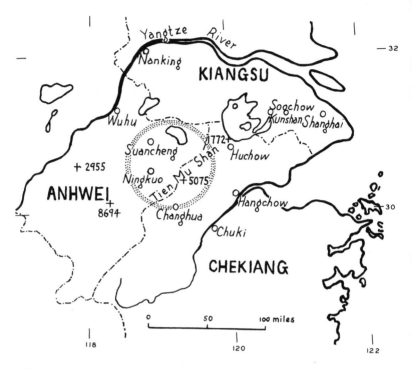

28. *Map of eastern China. Circle = approximate area of the original habitat of the ginkgo; elevations in feet.*

Henry and Barclay, apparently failed to consult. Their suggestion that the original ideographs concerned being *yin kuo* (silver fruit) is untenable as Kaempfer clearly indicates his ginkgo to be a transliteration of *yin hsing* (silver apricot). Moreover, the name *yin kuo* does not exist either in the literature or as a colloquial name in China and Japan. It is not found in the twenty-five such names for the tree and nut given by Moule (1937).

The name ginkgo appears in the fifth and last fascicle of Kaempfer's work, published twenty years after he left Japan. This fascicle lists and describes the economic plants of

Japan. He gives for most of the plants first their names in Chinese ideographs which, as he clearly states in the preface, are the literary names. These Chinese ideographs as pronounced in Japan are romanized and these are followed by romanized Japanese colloquial names. In a few cases the literary names are the same as colloquial names and these are indicated as such.

Thus in the case of ginkgo, the Chinese ideographs *yin hsing* are given at the beginning of the item (page 811) and again on the plate (page 813). Ginkgo is not "the scarcely known form" of Moule, but the literary name as occurring in Japanese literature at the time and especially in all herbals then in use, which were either translations from Chinese or Japanese works with plant names of Chinese origins.

Now that the derivation of the name ginkgo was clarified there remained the problem of the form of transliteration. The first line of the item in Kaempfer reads thus: "Ginkgo, vel Gin an, vulgò Itsjò. Arbor nucifera folio Adiantino." This clearly shows that Kaempfer considers that there are two ways of pronouncing the same ideographs, either as *"Ginkgo"* or *"Ginan."* It is also to be noted that in the description following he says that the fruit is called *"Ginnan."* The above-quoted lines of Kaempfer, including the Chinese ideographs, are cited in Endlicher's *Synopsis Coniferum* (1847) in the synonymy of *S. adiantifolia* Smith.[2]

As is well known, the same Chinese ideographs may have

---

[2] Endlicher gives in his monograph of the conifers a number of Chinese names in ideographs, either cited from Kaempfer or added by himself. Besides Kaempfer, he is one of the few European botanists to use Chinese characters in their publications. It may be mentioned that botanist Endlicher was also competent sinologist, although he did not seem to have studied the botanical aspects of the Chinese literature.

several different pronunciations in Japanese, varying from Japanese sounds to several types of Chinese sounds borrowed from China at different times and from different geographical areas. Kaempfer indicates two different ways of pronouncing ginkgo. In the Japanese herbal *Honzo Zufu* of 1828 the name "Ginkgo" is also given in two sounds, with the note that "Ginnan" is a Chinese pronunciation of later origin. Koidzumi (1936) and Makino (1951) quote other Japanese authors that *icho* is the Chinese pronunciation of *ya chio* in the Sung dynasty, and the former says also that *ginnan* is a Chinese pronunciation of later times. I believe both these pronunciations, *icho* and *ginnan,* came to Japan from southern Kiangsu in China, near the region where the tree originated. In the modern Wu dialect of the area, *ya chio* is pronounced *ai cho* and *yin hsing* as *nin an,* both closer to the Japanese names than to the Mandarin, or northern Chinese.

The most disputed fact is the seemingly quaint spelling "Ginkgo" of Kaempfer. Moule (1944) quotes Arthur Waley that "Ginkgo is a distortion of Ginkyo, which is the logical and regular reading of the Chinese name (Yin Hsing), but has no existence in the real language," adding the remark that this "is the true account of the word."

These authors, however, are considering the modern pronunciation and romanization of the word while Kaempfer was writing about his observations made over 250 years ago. The pronunciation of these words at that time might be slightly different. Fortune (1863), for instance, observing this plant in Japan about a hundred years ago, says that the nut is called *ging ko* in Japanese shops.

In the work of Kaempfer, the individual Chinese character *hsing* is romanized elsewhere as *kjoo* or *koo.* In the preface he says that some of the sounds, like *k* and *g,* especially in composite names, are not readily distinguished, and he gives examples as *K*oquan or *G*oquan and Kinari *g*aki or Kinari *k*aki. Thus in the case of ginkgo, he might be rendering a sound which he thinks is closest to the one then prevalent.

Kaempfer's romanization of this word may be a vagary, according to modern pronunciation, but the difficulty of explaining this, as Barclay (1944) says, "to any one conversant with the vagaries of the many systems used for 'romanizing' the Japanese sounds, . . . will certainly not seem insuperable." The fact to be noted is that Kaempfer was romanizing sounds of over 250 years ago and at a time when even the many different systems of romanizing Japanese sounds were not in existence. Thus the modern rendering *kyo* is to him *kjoo* or *koo* or *kgo,* the first two as individual sounds, and the last in combination with other sounds. Also the name *icho* was to him *itsjo.* The word "Ginkgo" is apparently not a misprint as suggested by Moule, as it is correctly listed also in the index of Kaempfer's work together with the other words discussed here. To reject his spelling because it does not conform with modern pronunciation seems most illogical. The name "Ginkgo" may be worth looking into etymologically, but as the original spelling of Kaempfer was correctly adopted by Linnaeus as a botanical name, nomenclaturally speaking, there is no ground to reject it as an "unintentional orthographic error." As ginkgo it has been known for nearly 250 years and certainly will be known as such forever.

5

The Horse Chestnuts

Among shade trees, horse chestnuts and buckeyes have a special appeal because of their very striking foliage and flowers. The large, palmately compound leaves are unique among trees of the temperate regions. The long, upright clusters of colorful flowers are showy and conspicuous. These, together with their stately, compact, pyramidal shape make them among the most popular ornamental trees in all continents of the Northern Hemisphere.

Most people consider the trees to be magnificent in both foliage and flowers. Wilson (1920), for instance, speaks enthusiastically of the common horse chestnut as the most handsome exotic flowering tree in the eastern United States. But dissenting views are sometimes also expressed, such as by Wyman (1951), who considers the tree coarse in foliage texture, flower, and fruit and a "dirty" one, as it is always dropping something.

THE GENUS AESCULUS

The horse chestnuts and buckeyes belong to the genus *Aesculus,* of the Hippocastanaceae, or horse chestnut family, a genus of some fifteen species of trees or shrubs of the temperate regions of the Northern Hemisphere. The leaves are oppositely arranged, each with five to nine digitately disposed leaflets. The flowers are four- or five-petaled, white or showy-colored, and are borne in upright, many-flowered clusters. The fruit is large, rounded, either smooth or prickled outside, splitting into three valves exposing the single large seed. The seed or nut is unfit for human consumption.

Native species are cultivated in Europe and Asia and also in North America. The bringing together of European and American species in cultivation, and the several different American species themselves, have resulted in the production of many hybrids, some superior to the parental species in certain qualities. The very large number of hybrids which have arisen in cultivation within the last 200 years renders this group a distinctive one in the genetics of trees and ornamentals. The significance of hybridity in this genus is further enhanced by the occurrence of chromosome doubling in certain forms which renders these fertile and true-breeding and leads to the production of hybrids with intermediate chromosome numbers between the low- and high-chromosome types. The story of hybridization in the genus *Aesculus* is documented by cytological studies and well illustrates what can be done in deliberate breeding of trees in many families and genera.

EUROPEAN SPECIES

The common horse chestnut, *A. hippocastanum,* native to the Balkan peninsula, is the only European species of the genus. It is a large tree, attaining a height of 100 feet, with five- to seven-foliolate leaves, sessile leaflets, white flowers tinged with red, and prickly fruit. It is the first of all shade trees to burst into leaf, and it is desirable not only as a street tree but also as a specimen tree in parks or private grounds.

In Europe and in England it is now widely planted and has become such a common feature of the landscape that most people assume that it is native. In England, the most famous planting is the avenue leading to the Hampton Court Place at Bushey Park on the bank of the Thames, a mile in length and 170 feet wide, and lined with 137 trees on each side set 42 feet apart. It was planted by the famous architect Sir Christopher Wren in 1699. The tallest trees are over 100 feet high and the show of blossoms is truly spectacular (Wilson, 1920).

The native home of this species was long an enigma to botanists. Linnaeus considered the tree indigenous to northern Asia and De Candolle thought that it came from northern India. Only Decaisne expressed the opinion that the plant was of European origin, and this was confirmed and the enigma finally solved by the discovery made by Orphanides of the tree growing in a wild state in the mainland of Greece (in a note in the French translation of Griesbach's *Die Vegetation der Erde,* made by M. de Tchihatschef, 1876). It is now definitely established that its native habitat is confined to northern Greece and Albania. Its occurrence in other regions is in all cases due to cultivation.

It is uncertain whether or not the horse chestnut was known to the ancient Greeks. The earliest record of cultivation is in Constantinople. The fruit was known as *at-kastan* (horse chestnut) to the Turks, who found them useful as a drug for horses suffering from broken winds or coughs (Loudon, 1838; Wilson, 1920).

Seeds of the horse chestnut were sent in 1570 from Constantinople to Vienna by Dr. von Ungnard, Imperial Ambassador to the court of Suleiman II, and a tree was raised by the celebrated botanist Clusius. Seeds were brought to France from the Levant by Bachelier in 1615, and the tree was probably introduced into England at about the same time (Loudon, 1830).

Although the exact dates of many plants introduced into this country by early English settlers are so often unknown, fortunately, as noted by Wilson (1920), the introduction of the common horse chestnut into America is a matter of record. The exact date is noted in the letters published by William Darlington in his *Memorials of John Bartram and Humphrey Marshall,* in 1849. Seeds were first sent by Peter Collinson in 1741 from London to John Bartram in Philadelphia, and in 1763 Collinson learned of the flowering of his horse chestnut. Collinson spoke also of the long horse chestnut avenue at Hampton Court (the trees were then 50 feet high), as "one of the grandest and most charming sights in the world." In recording this introduction, Wilson especially remarked that, based on his own experience, he considered the horse chestnut among the most difficult to transport safely and he marveled that in those days of slow sailing ships its introduction into the New World should have been successfully accomplished.

The common horse chestnut is now widely planted and there are numerous garden forms which have originated in cultivation. There is a double form 'Flore-pleno.' In leaf variations, there is 'Memmingeri' with yellow leaves, 'Aureovariegata' with yellow variegated leaves, 'Laciniata' with narrowly lobed leaflets, and 'Crispa' with stalked and broad leaflets. In habit 'Pyramidalis' is upright, 'Umbraculifera' is dense and globular, and 'Tortuosa' has bent and twisted branches. All these garden forms are propagated by cuttings or grafting and are thus to be considered as clones.

HIMALAYAN AND SOUTHERN ASIATIC SPECIES

There are two species of *Aesculus* native to the Himalayan region and southern mainland Asia. One of them, the Indian horse chestunt, *A. indica,* a large tree growing 150 feet high, is common in the northwestern Himalayas from the Indus to Nepal, and also in Afghanistan. It occurs at elevations from 4,000 to 10,000 feet. The seed is given as food to cattle and goats, and, ground and mixed with ordinary flour, is also part of the diet of the hill tribes (Elwes & Henry, 1907). The tree is occasionally cultivated in England but does not seem to have been introduced into North America.

Another species, *A. punduana,* occurs in northeastern India, in the Sikkim and Khasia Hills through Upper Burma and Thailand to Tonkin in Indo-China. It is a large tree with very large leaflets. Essentially a subtropical tree, is is not cultivated elsewhere.

EASTERN ASIATIC SPECIES

Three species occur in eastern Asia: two in China and one in Japan, all of them occasionally known in cultivation in other temperate regions. Although the three species are distinct both morphologically and geographically, they have been much confused with each other in botanical as well as prescientific writings.

In northern and eastern China, the Chinese horse chestnut, *A. chinensis,* is widely grown as a shade and ornamental tree, especially on the grounds of Buddhist and Taoist temples. The tree is probably native to Shansi and Shensi in northern China and the present range is much expanded because of cultivation.

The Chinese horse chestnut is a medium-sized tree attaining a height of 75 feet. The leaves are composed of five to seven large and stalked leaflets. The flowers are white and small, forming panicles about 8 inches long. The generally globular fruit is about 3/4 in. across, more or less depressed above, and covered with warts but is not spiny.

In western China, in Szechuan and Hupeh, there occurs a horse chestnut originally regarded as belonging also to *A. chinensis,* but Rehder (1913) separated it as a distinct species, *A. wilsonii.* It is very similar to *A. chinensis* in general appearance, differing mainly in its hairy and longer-stalked leaflets, in its larger flowers, and in the more rounded fruit with larger seeds. It is a tree of the mountains, occurring in woods at elevations of 3,000 to 6,000 feet.

In Japan, there is found *A. turbinata,* which was formerly much confounded with the Chinese horse chestnut. The Japanese horse chestnut has stalkless instead of stalked leaflets and five instead of four petals. The flower is yellowish-white and the fruit is broadly pear-shaped. This tree is cultivated in Japan and also in some cities in eastern China.

In China *Aesculus* is often called "seven-leaved tree," referring to the palmately compound leaves, or sometimes as "monkey chestnut" because of its original mountainous habitat and of the fruit unfit for human consumption.

A special significance is attached to the horse chestnut in China and Japan by the Taoists and Buddhists. In the hills around Peking, and in many other parts of China, the tree, which the Buddhist monks call *so-lo* is frequently planted on temple grounds. This name is a transliteration of the Sanskrit *sala,* which originally refers to *Shorea robusta* of India, one of the sacred trees of Buddhism. In a grove with a mighty *sala* tree, the Buddha Gautama, founder of Buddhism, was born, his mother clutching a branch of the mighty tree. An even greater sanctity was given to this species because Gautama on his last mission died, lying on his cloak, which his disciple Anada folded and placed for him on the ground between two *sala* trees (Burkill, 1946).

As *Shorea* is not hardy in northern China, Buddhists anxious to grow the tree about their establishments transferred the name to *Aesculus,* an attempt, as Burkill says, made in a kind of desperation. Burkill thinks that it is remotely possible that this substitution began in Kashmir monasteries when, in their inability to grow *Shorea,* the monks substituted the Indian horse chestnut, *A. indica,* but he says this has not been demonstrated. Thus, Burkill asserts that "The Chinese, whose pilgrims were certainly familiar with Kashmir, for some unexplained reason took their *Aesculus chinensis* Bunge, which is very like *A. indica,* for 'Sala,' calling it 'So-lo,' planting it about their monasteries."

However, the substitution of the name is entirely of Chinese origin, and as noted by Burkill, it is not concerned with the Kashmir plant. The reason for choosing this plant, which puzzles Burkhill, is to be found in the religious significance attached to the plant in China prior to its adoption by the Buddhists, apparently only since the sixteenth century.

In China, *Aesculus* is also known by the name *tien-shi-li,* or "Chestnut of the Heavenly Teacher"; this name appears first in the book, *Account of the Products of Szechuan,* by Sung Chi of the eleventh century. It states that the tree is found only in the Ching-chen mountains of Szechuan and was reputed to be planted there by Tien-shi, the Heavenly Teacher, referring to the Taoist pope, Chang Tao-ling of the first century A.D., when he studied and meditated there, seeking for *tao,* or the truth. From the locality, apparently the name refers more specifically to *A. wilsonii.*

Li Shi-chen, in his great herbal *Pen Ts'ao Kang Mu* of 1558, identifies this *tien-shi-li* with the plants called *so-lo,* a name which he says was unknown in former times. Thus it is clear that the Buddhists adopted this plant almost five centuries after the Taoists attributed any significance to it.

Along with these attributions, the seeds of *Aesculus* were reported to have benign influence in exorcising demons. The transference of the sacred Buddhist tree *sala* from *Shorea* of India to *Aesculus* in northern China, out of necessity because of climatic incompatibility, is thus apparently motivated by its original religious and superstitious significance.

AMERICAN SPECIES

There are no less than ten species of *Aesculus* in North America, which thus has the largest concentration of the genus. These species, generally known as "buckeyes" in America, are usually smaller trees or shrubs. They have mostly five leaflets. Among the larger species, trees or sometimes large shrubs, there are about seven species. There are two species in California, *A. parryi* and *A. californica.* The latter, the California buckeye, the more common species, is a tree 35 feet high and with white to rose-colored, fragrant flowers. It is in cultivation but suitable only for the southern states.

The largest American species is the sweet buckeye, *A. octandra* of eastern North America. It grows to a height of 90 feet, and bears flowers which are usually yellow. Another large tree is *A. sylvatica* (*A. neglecta*), native from North Carolina to Alabama, which attains a height of 60 feet, with pale-yellow flowers.

Among the smaller trees in the Ohio buckeye, *A. glabra,* state tree of Ohio, a tree to 50 feet tall with relatively small, bright-red flowers (Figure 29). Another shrub or small tree of the same stature is *A. discolor* of the southern states, which bears yellow flowers flushed with red.

29. Aesculus glabra, (Photo, Ernst Schreiner).

30. Aesculus parviflora. (Photo, Ernst Schreiner.)

31. Aesculus parviflora, flower clusters. (Photo, Ernst Schreiner).

Among the species mentioned above, the most commonly planted are the Ohio buckeye and the sweet buckeye. These two species have also been introduced into England and the European continent and are frequently planted there.

The remaining species are mostly shrubby. Among them are some valuable ornamentals for the lawn, such as *A. parviflora* (Figures 30, 31) and *A. splendens,* of the southeastern and the southern states, respectively. The latter is a spreading shrub readily propagated by root cuttings, and is especially valuable for its autumn flowers of a very beautiful red color.

Aesculus parviflora, which is known as the "bottle-brush buckeye," is a shrub 3 to 10 feet tall, with white flowers and long exserted stamens. The plant spreads rapidly by horizontal runners and a single specimen may, in a few years, cover an area of many square yards.

HYBRIDS

The number of hybrids or putative hybrids is, in proportion to the number of species, exceedingly large in the genus *Aesculus*. The exact origins of most such hybrids are not known. Some occurred in the wild, but the majority originated in cultivation.

The common horse chestnut of the Balkans was first introduced into France and England in the first half of the seventeenth century and into North America in the eighteenth century. The American species have been in cultivation since the latter part of the eighteenth century. As early as 1764 the American sweet buckeye, *A. octandra,* was introduced into England, and the Ohio buckeye, *A. glabra,* was introduced in 1812 (Loudon, 1838). Many species of varieties of American origin, however, did not occur in cultivation until the beginning of the present century. Thus, the numerous hybrids of the genus could not have appeared in cultivation, either in this country and in Europe, before the latter part of the eighteenth century, and actually many of the hybrids were noted only in the last fifty years.

A large number of hybrids are recorded as crosses among the American species, but the most celebrated hybrid is the red horse chestnut, *A.* × *carnea,* supposedly a cross between the European *A. hippocastanum* and the American *A. pavia.* This plant must have originated before 1818, the year Loiseleur, who first described the plant as *A. rubicunda,* received it from Germany, but as there are no earlier accounts of it, the exact origin is unknown. However, it was long considered as undoubtedly a chance hybrid between *A. hippocastanum* and *A. pavia,* as it possesses characters of both the supposed parents. Among the characters derived from the former are the leaves and the slightly spiny fruits. Characters derived from the latter are the color of the petals and the glands on their margins. The red horse chestnut is a small tree usually about 30 feet in height. The flowers are red, with an orange blotch at the base of the petals which eventually become deep red.

Unlike most other hybrids, this plant breeds true from seed. This peculiarity, together with its hybrid nature, was later revealed by cytological studies. Hoar (1927) discovered that *A.* × *carnea* has eighty chromosomes while all the other species of the genus, including the two supposed parents, have forty. Thus this plant is a tetraploid, a plant with a double set of the normal diploid chromosome apparatus.

Aesculus × *carnea* was therefore presumably derived by the doubling of a sterile hybrid, which explains its capacity for breeding true from seed. This doubling of chromosome number was confirmed by Upcott (1936), who also found the chromosomes of the two parent species exactly similar in size and shape. These parent species, having forty chromosomes, show secondary pairing and the formation of an occasional quadrivalent, a phenomenon suggesting that the parent species themselves are tetraploids and that the normal chromosome number of the genus is twenty. The hybrid *A.* × *carnea* must therefore be regarded as an octoploid, a plant having a double chromosomal apparatus of a tetraploid.

The leaves of *A.* × *carnea* are of darker green than ordinary, of firmer texture and with an uneven surface, and the short stalks of the leaflets are more or less curved and twisted. These characters are attributable to the polyploidal nature of the plant, a plant having more than the normal diploid number of chromosomes. Its ability to resist drought better than the parent species is apparently also due to its hybrid nature.

The most interesting fact about hybridity in *Aesculus* is the occurrence of another hybrid, *A. × plantierensis,* a chance hybrid supposedly between *A. hippocastanum* and *A. × carnea.* This plant was raised in the nurseries of Messrs. Simon-Louis Frères at Plantières near Metz, France, from a seed of *A. hippocastanum* (André, 1894). It is intermediate in characters between the two supposed parents. It has whitish flowers suffused with pink and fading to pink.

This supposition that *A. × plantierensis* is a backcross of an octoploid progeny to one of the tetraploid parents was later confirmed by Upcott (1936). He found the chromosome number to be sixty, and therefore the plant is to be regarded as a hexaploid. This plant is sterile, with its chromosomes consisting of varying numbers of univalents, bivalents, and multivalents.

This plant is also considered as a variety of *A. × carnea* as "var. *plantierensis*" (Rehder 1940). As its hybrid nature is definitely established, it will be more appropriately designated as *A. × plantierensis,* the name originally proposed for it by André.

The relationship and parentage of the several plants in consideration illustrating as it does a most unique and striking case of hybridization, backcrossing, and chromosome doubling in trees, can be summarized by the following diagram:

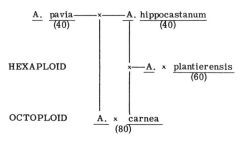

Among the numerous hybrids of garden origins between the American species, the most notable is the hybrid buckeye, *A.* × *hybrida,* a hybrid between *A. octandra* and *A pavia,* which originated before 1815. Its characteristics are intermediate between the two parental species. There are several forms of this hybrid in cultivation, with the color of the flowers varying from yellow slightly tinged with pink to nearly red throughout.

Many other hybrids have been cultivated since the beginning of the present century from plants which originated either in the garden or in the wild. In Rehder's *Manual* (1940), there are listed fifteen species of cultivated *Aesculus* and no less than sixteen hybrids or supposed hybrids of known as well as unknown origins.

6

The Lindens

The lindens, or basswoods, belong to the genus *Tilia,* a member of the linden family, or Tiliaceae. Although this family is essentially tropical, *Tilia* itself is confined to the temperate regions of the Northern Hemisphere and is represented by some thirty species in Europe, Asia, and North America. All the species are trees with alternate, simple, usually cordate leaves and flowers and fruit clusters borne on large leaflike bracts, the latter being a very unusual feature among flowering plants.

The lindens are handsome trees, with beautiful foliage, and are widely planted as ornamental or shade trees. The continuous falling of the fruit bracts, however, make them somewhat untidy. Besides being popular ornamental trees, the linden possesses several characteristics which have been of great value to mankind since very ancient times. One is their importance as a source of honey, for their flowers are very fragrant and rich in nectar. Another is the use of the wood for carving and the making of utensils, as it is soft, straight-grained, and easily worked. Still another is the utility of the inner bark, or bast, which is fibrous and can be used for the manufacture of cord or fishing nets.

The value of the soft wood in making carvings and the bast for making cord was known not only to the ancient Europeans and Asians, each employing separately their own native species, but was also discovered independently by the American Indians, who utilized the native American species. The various uses in the different lands will be discussed below in greater detail.

COMMON NAMES

The common names of *Tilia* are many and varied. In the United States, the various species are usually called linden or basswood (bastwood), but sometimes the trees are also known as "bee tree" or in the South as "linn." The wood, because of its pale color, is commonly known in America as "whitewood," a name also applied to the wood of the tulip tree, *Liriodendron*. The most commonly used name of the trees in Europe is "lime," probably derived from the old Anglo-Saxon name *linde*. *Linde* is the common name used by the Germans and the Dutch. The classical Roman name of the tree was *tilia*. This becomes *tilleul* in French, *tiglio* in Italian, and *tilio* in Spanish. Linden is an especial favorite of the Germans and is often mentioned in their folklore and early poetry.

In eastern Asia the classical name given to the trees by the ancient Chinese is *tuan*. Many colloquial names are locally applied to the various species.

EUROPEAN SPECIES

There are three species of *Tilia* commonly cultivated in most countries of Europe (Figure 30). Two of the species have been in cultivation since very ancient times: *T. platyphyllos,* the large-leaved linden, with large, hairy leaves and large flower clusters and *T. cordata,* the small-leaved linden, with smaller and hairless leaves and smaller flower clusters. A third, *T. europaea,* the common linden, now more commonly planted than the other two species, has characters intermediate between the two and is in all probability a hybrid between them (Figure 32).

The lindens were known to the Greeks and Romans and were mentioned by Theophrastus, Pliny, and many other early writers. The use of the tree as a honey source made it a favorite in the ancient world. The tree, according to Theophrastus, is of two sexes, totally different in forms, referring probably to the large-leaved and small-leaved species (Loudon, 1838). It was much esteemed by the Romans as a shade tree. According to Pliny, the wood was employed for numerous uses. Virgil mentions especially the quality of yokes made from its wood.

The early classification of the European species was somewhat confusing, and a very long list of synonyms is involved. The three lindens were recognized by Linnaeus in 1753 as representing a single species, *T. europaea,* with eight different varieties.

The large-leaved linden, *T. platyphyllos,* is widely distributed through central and southern Europe. It has been in cultivation longer and has produced more varieties than the other species, among them fastigiate and laciniate-leaved forms.

The small-leaved linden, *T. cordata,* a less variable species, is the common linden of northern Europe, although it is also found in the south especially at high elevations. Though it is common and widely distributed, it is less frequently planted in central Europe than the other lindens.

The common linden, *T. europaea* (*T. vulgaris*), is the largest of European lindens. As mentioned before, it has characters intermediate between the two other species. Its hybrid origin is further indicated by the fact that although it is the most widely planted species in Europe, it is almost unknown in the wild (Sargent, 1889; Elwes & Henry, 1912). Its probable hybrid origin was early recognized by such authors as Palmstruch and Venus, who called it *T. intermedia* in 1802, and Bechstein, who called it *T. hybrida* in 1810. In 1909, in his monograph of the genus *Tilia,* V. Engler definitely identified it as a hybrid between *T. platyphyllos* and *T. cordata* (Figure 32).

Just when the common linden first appeared in cultivation cannot be ascertained. It has been cultivated as a park and avenue tree for at least three and a half centuries (Sargent, 1889). In the early years it was largely propagated in Dutch nurseries and was often known as *"Tilia hollandica"* or *"Hollandische Linde."* It is the tree that gives its name to the most famous street in Berlin. This is also the lime so often seen in English parks and gardens. The most famous avenue there is the one in Trinity College at Cambridge.

Besides these three lindens which are widely planted in Europe, there are three other species: *T. tomentosa, T. petiolaris,* and *T. dasystyla,* which grow in the southeastern part of the continent as well as in western Asia. *T. tomentosa,* the silver linden, is a tall ornamental tree with silvery leaves (Figure 33). *T. petiolaris,* when mature, is a large tree with pendent lower branches; its leaves are whitish-tomentose on the lower surfaces.

USES OF LINDEN IN EUROPE

Lindens have been valued since ancient times for their honey. The honey produced from the flowers is universally considered superior to all other kinds in quality. Linden oil, obtained by distilling the flowers, has a pleasant odor and is used in perfumery.

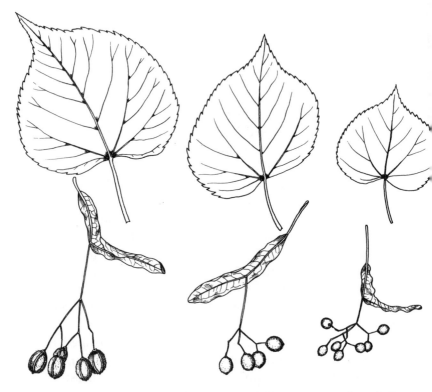

32. Leaves and fruiting clusters of Tilia. Left: T. platyphyllos; middle: T. europaea; right: T. cordata.

The wood of linden is known for its easy working quality, and besides many other uses, is highly prized for carving. Many of the fine carvings in old buildings in England as well as on the European continent are of this wood. In ancient times, it was extensively employed for making all kinds of utensils.

The most important value of lindens, especially in the northern part of Europe, was for the making of ropes, mats, shoes, fishing nets, and similar articles from the inner bark, or bast. This use was of great importance to peasants in

33. Tilia tomentosa.

former times, but even today ropes and mats are still being made from linden bast in some parts of northern Europe.

Lindens have long been considered as among the most favorite avenue trees in Europe. The fashion of planting these was made popular by French landscape gardeners in the later half of the seventeenth century, at a period when the formal style of gardening prevailed. Avenues of linden trees were an essential feature in every park and town in central and northern Europe. The ability of lindens to thrive under repeated severe pruning makes them especially fit for formal gardens. Their graceful, free habit renders them desirable as a lawn or street tree.

Mention may be made here of the "honey dew," somewhat viscid, sugary dots frequently found sprinkled on the upper leaf surface of the common linden and related species. Much has been written on this subject. Some authors believe this is an excretion caused by the presence of aphids, but most probably it is a direct exudation from the leaves due to excessive transpiration under intense sunlight (Elwes & Henry, 1913). After the honey dew dries and thickens, it becomes the site of growth of certain fungi, resulting in the blackened dots commonly present on the leaves in late summer and fall.

EASTERN ASIATIC SPECIES

Over two-thirds of the species of the genus *Tilia* grow in eastern Asia. There are about twenty-one species in China, one in Korea, and three in Japan. The species are mostly tall trees of the mountain forests, and only a few are in cultivation in their native countries. The most widely cultivated species in northern China is *T. mandshurica,* the Manchurian linden. The most frequently grown species in eastern China is *T. miqueliana,* which is also commonly planted in Japan. A few other species are occasionally planted, including *T. tuan* of central China, the specific epithet given by the Polish botanist Szyszylowicz being the classical Chinese name of linden trees.

In northern China, the species of *Tilia* are considered valuable not only as shade and ornamental trees but also for their wood and bark. The wood is used for manufacturing various kinds of utensils, for carving, and for making printing blocks. The bark is used extensively for making ropes and other coarse woven products such as mats and sandals. Large quantities of the fiber are collected by the farmers in northern China for these purposes.

Many of the wild species of *Tilia* have been introduced into cultivation in the West by modern explorers, but these are still limited to botanical gardens and arboretums.

LINDENS AS BUDDHISTS' TREES

An interesting feature of lindens in eastern Asia is the religious significance attached to them by the Buddhist monks. This is not an ancient attribute, but one which was deliberately accorded to it by these priests perhaps some five or six hundred years ago, after Buddhism, introduced from India several hundred years earlier, had become firmly established in China and very popular among the general public.

The Buddhists in China call the linden *p'u t'i shu,* or *bodhi* tree, a name originally applied to *Ficus religiosa,* a tropical tree. According to the Buddhist legend it was under a tree of *F. religiosa* that understanding came to the Buddha Gautama, in 528 B.C., after eight years of meditation. The tree was named *bohdidruma* in Sanskrit, meaning "the tree of enlightenment" (Burkhill, 1946). In latter years cuttings from this sacred tree were taken and planted in different parts of India and Ceylon, and from these trees cuttings were also brought to southern China.

In northern China, however, this tropical tree is not hardy. In the words of Burkhill, the Buddhists "in a kind of desperation," transferred the name to *Tilia.* He thinks there is no similarity between the two to justify such a transfer. However, I believe the basis for such a transfer may be in part due to the slight resemblance in the leaves, and in part due to the unique fruit clusters of the lindens. The leaves of both are broad at the base and with a cuspidate tip, giving them more or less a similar outline. The fruits of lindens are of course very different from those of *Ficus,* but they are distinctive in appearance, resembling somewhat the fruits of another plant, *Sapindus mukorossi,* which bears fruits long attributed by the Taoist priests as possessing exorcizing power. This was called *p'u t'i tze,* or *bodhi* fruit by the Buddhists in later times (Chen, 1946). Thus the transference of the name *"bodhi* tree" is made actually to two plants, one for the leaves and another for the fruit.

The species of *Tilia* cultivated by the Buddhists on temple grounds as *bodhi* trees are two: *T. mandshurica* in northern China and *T. miqueliana* in eastern China. The latter is rarely found in the wild and is quite variable, thus suggesting a possible hybrid origin. It was introduced into Japan and is widely planted there, mostly on temple grounds. Although Japan has several native species of *Tilia,* this introduced species is the one commonly cultivated. It bears the name *bodai,* Japanese for *bodhi* (Makino, 1951), and is traditionally believed to have been introduced from China by Buddhist priest about the year 1190 A.D. (Elwes & Henry, 1912).

AMERICAN SPECIES

There are about five species in North America. The most widespread is *T. americana,* the American linden, which has a natural range in the eastern part of the continent extending from Canada in the north, southward to Alabama, and westward to Texas and North Dakota (Sargent, 1891). It is the most commonly cultivated native species, frequently planted as an avenue tree (Figure 34).

The American linden is one of the earliest cultivated trees in North America. Its first cultivation must have been in the first half of the eighteenth century. Rehder (1940) gives the date "Intr. 1752," which is misleading. This is the date of cultivation by Miller in England (Loudon, 1838). Actually its first introduction into England was much earlier, for in the *London Catalogue of Trees,* 1730, there is mentioned: "Tilia, with leaf more longly mucronate. Seeds sent from Carolina by Catesby in 1726, hardy, and may be propagated as other limes" (Elwes & Henry, 1913).

34. Tilia americana.

In Europe, several varieties with different leaf forms have been selected from cultivated plants: 'Ampelophylla,' with large coarsely and irregularly dentate leaves; 'Dentata,' with more sharply, often nearly double dentate leaves; and 'Macrophylla,' with larger leaves than the typical form. In America, there is now in cultivation, a pyramidal form 'Fastigiata,' with ascending branches.

35. Tilia heterophylla.

The other native American species are less commonly cultivated. *T. heterophylla* (Figure 35), together with the var. *michauxii*, is also widely distributed in the eastern United States. The other species are more restricted in their ranges, *T. neglecta* of eastern North America, *T. monticola* of the Appalachian Mountains, and *T. floridana* of the southern United States. Incidentally it may be mentioned that there are widely different views among botanists regarding the classification of American lindens. Some recognize only three species, and others as many as fifteen.

USES OF LINDENS IN AMERICA

The American Indians, before the arrival of Europeans, independently made use of the special qualities of lindens, namely, the soft, easily worked wood, and the tough fibrous inner bark, or bast. The bast was used for making cord and matting. The wood was much used for making household wares. Longfellow narrates that all the bowls at Hiawatha's wedding were made of basswood, smoothly polished.

The wood, known also as "whitewood," is light brown in color. It is now used in America in the manufacturing of wooden ware, cheap furniture, and similar objects. It is also used for making paper pulp, but the quick decomposition of the sap make it unfit for white paper (Sargent, 1891).

EUROPEAN SPECIES IN AMERICA

The common linden and some other species of Europe are often cultivated in America, either as avenue trees or in botanical gardens. The American species, in fact, do better than European trees, since they are native and therefore more suitable to the climate. But European lindens are still being widely planted as street or lawn trees in this country.

Many of the larger eastern cities in North America have a "Linden Avenue" and often it is the European common linden that lines the street. These European lindens were doubtless among the very first trees brought to the New World by the early colonists. There are, for instance, massive trees of the common linden planted along the streets at Plymouth, Massachusetts.

Lindens are subject to the attacks of many insects, which sometimes disfigure the trees by devouring the foliage or even destroy the trees by boring into the trunks. All the American and European species are susceptible to such injuries, but trees growing in richer soils often fare better and are less affected than native species.

HYBRIDS

It was mentioned above that the common linden (*T. europaea*) is most probably a hybrid between the two common European species. The species in this genus seem to be able to hybridize freely. With the introduction of American trees into Europe, there have arisen in cultivation a number of specimens showing intermediate characters between some of the American species and the various European ones. Hundreds of these putative hybrids have been described. A list of the better known ones is given below; their parentage is often conjectural. These were all selected from among cultivated plants in Europe in the nineteenth century, and most of them are not in cultivation in America.

T. × euchlora (*T. cordata × dasystyla*?) Originated before 1860 in Russia.

T. × *flaccida* (*T. americana* × *platyphyllos?*) Originated in France before 1834.

T. × *flavescens* (*T. americana* × *cordata?*) Originated before 1836 in Germany.

T. × *juranyana* (*T. cordata* × *tomentosa?*) Originated before 1834 in Hungary.

T. × *moltkei* (*T. americana* × *petiolaris?*) Originated before 1880 in Germany.

T. × *orbicularis* (*T. euchlora* × *petiolaris?*) Originated about 1870 in France.

7

The Maples

The brilliant display of colors in the forests and country-side every autumn is a familiar sight to people living in the northeastern part of North America. Few realize, however, that this autumn splendor is restricted to only certain parts of the world. Such colorful scenery is not seen by residents in the tropics, nor in other parts of the temperate world, such as western North America, where the temperature does not fluctuate widely throughout the year. In fact this gay mantle of fall color may be seen only in temperate eastern Asia, eastern North America, and parts of central Europe.

In these three areas, the monthly temperature change is pronounced and a more or less severe winter follows a warm summer. At the approach of winter, the deciduous trees prepare themselves for a long resting period. Foods manufactured by the leaves in the form of sugars are trans-ferred to the woody branches or roots for storage, chiefly in the form of starch, for the growing buds next spring. During this process, chemical substances known as antho-cyanins are produced in the leaves of many plants. They may appear red when free acids are present in the cell sap and blue when there are no acids.

The red color of the autumn leaves is due to these pig-ments. The color is more intense in some trees than in others. If anthocyanins are associated with yellow compounds, the leaves will assume an orange tinge. The yellow color is due to the yellow pigments normally present in the chloroplasts, the green granules which manufacture food for the plant in the sunlight. In the dying leaves, the green pigment of the chloroplasts disappears and the yellow color thus becomes visible. Eventually the leaves will dry and turn brown. Thus, in autumn coloration, plant foliage displays all gradations from brown to yellow to red.

Among the trees that show the most brilliant color in autumn are many maples, sweet gum (*Liquidambar*), sassafras, tupelo (*Nyssa*), scarlet oak, sorrel tree (*Oxydendrum*), and also some dogwoods, sumacs, viburnums, and other genera. Most of these genera are especially developed in, or restricted in their distribution to, eastern Asia and eastern North America. Because of its large number of species and their wide distribution the most important genus of all is *Acer,* the maples.

ACER, THE MAPLE GENUS

The maples are among the most widely planted shade and ornamental trees. They are used everywhere, along streets, in parks, and in gardens. The genus contains trees of such sizes and types as to meet all requirements, and the great variation in size and form of the species and in the shape and color of the leaves is not matched by any other genus. These characters, together with ease in transplanting and their relative freedom from serious pests or diseases, make these trees highly popular for landscaping purposes.

This large genus contains at least 150 species, distributed over a major part of the Northern Hemisphere from Alaska and Siberia southwards. The genus also occurs south of the equator in the islands of southern Asia—in Sumatra, Java, Celebes, and Timor (Mulligan, 1958).

The majority of the species occurs in eastern to southeastern Asia, with sixty-nine in China, sixteen in Japan, and nine in southeastern Asia. There are about seven species in India and Burma to Tibet, seven in western Asia, and thirteen in Europe. In North America, including Mexico and Guatemala, there are altogether twenty-five species.

About sixty-eight species of *Acer,* with numerous varieties and forms, are in cultivation in North America. Their most important use is as street trees. Other species are used in rock gardens or on lawns. The autumn coloration of the foliage of many species is the most desirable feature for landscaping. Some species also provide colorful foliage for the spring or summer. A number of kinds are valued also for their ornamental bark (Wyman, 1959).

MAPLE FOR AUTUMN COLORATION

Many species of maples are cultivated primarily for their autumn color. This color varies from red to red and yellow, orange-brown, and yellow. The commonly cultivated species with these different colorations are listed below:

Red:

A. capillipes	*A. mandshuricum*
A. ginnala	*A. nikoense*
A. griseum	*A. palmatum,* many forms
A. japonicum	*A. rufinerve*
A. leucoderme	*A. tataricum* 'Rubrum'

Red and/or orange and red:

A. buergerianum	*A. rubrum*
A. circinatum	*A. saccharum*
A. cissifolium	*A. spicatum*

Orange or orange-brown:

A. carpinifolium	*A. palmatum,* some forms
A. macrophyllum	*A. truncatum*

Yellow:

A. campestre
A. cappadocicum
A. glabrum
A. mono
A. nigrum
A. pensylvanicum

A. platanoides
A. saccharinum
A. tataricum
A. tegmentosum
A. tschonoskii

Yellow and purple:
A. davidii

SUMMER COLORATION

Besides early spring and late autumn, some maples also provide foliage of varied colors in the normal growing season from late spring through early autumn. A few species are noted for the brightness of their green color, but it is the large number of distinctive colored and variegated forms that are valued for specific ornamental purposes. The greatest variation in color is found in the Japanese maple, *A. palmatum,* which we will discuss separately. The following is a list of colored and variegated forms among the other species of the genus.

Colored forms:

A. heldreichii 'Purpuratum'	Red beneath
A. japonicum 'Aureum'	Yellow
A. negundo 'Auratum'	Yellow
A. platanoides 'Rubrum'	Dark red
'Schwedleri'	Red beneath
A. pseudo-platanus	
'Purpureum'	Purple beneath
'Worleei'	Deep yellow
A. velutinum 'Wolfii'	Red beneath

Variegated forms:

A. campestre 'Variegatum'	Bordered with white

'Albo-variegatum'	With large white blotches
A. negundo 'Variegatum'	Bordered with white
'Aureo-marginatum'	Bordered with yellow
'Aureo-variegatum'	Spotted yellow
A. platanoides 'Variegatum'	With large white blotches
A. pseudo-platanus	
'Flavo-variegatum'	With yellow blotches
'Variegatum'	With white blotches
A. rufinerve 'Albo-limbatum'	Bordered with white dots

THE JAPANESE MAPLE

The above lists show the great variation in color of the foliage within the genus. Besides color, variation in leaf form in *Acer* is also very great. The leaves vary from un-

36. Acer palmatum.

37. *Acer palmatum, a dwarf, dissected-leaved form.*

lobed to lobed and finally dissected forms (Figures 36, 37).

All these variations, in color as well as form, are highly manifested in a single species, *A. palmatum.* The Japanese maple is undoubtedly the most variable species, so far as foliage is concerned, of cultivated trees or shrubs. Mulligan (1958) recognizes more than seventy varieties or cultivars of *A. palmatum* in North America. While in other ornamental plants, especially in herbaceous ones, variation frequently occurs in flowers, here the ornamental feature depends mainly on the leaves, and sometimes also on the shape of the plant.

This great variation is brought out by intensive cultivation and selection in the Japanese garden. The species has been cultivated there since very early times for the brilliant red foliage in autumn so frequently praised in poetry and depicted in paintings. The Japanese call it " Takao maple" because it is especially abundant on the mountain Takao, famous since ancient times for autumn coloration. They use it extensively in their gardens and also as a potted dwarf tree (Makino, 1951).

The Japanese maple is a shrub or small tree. It is native to Japan and adjacent parts of the Asiatic mainland. In the Japanese literature there are hundreds of named forms, many of which are now also in cultivation in Western gardens. The variation may be either in the color or the shape of the leaves or sometimes in a combination of these two characters.

The leaves vary from five- to nine-lobed. In some forms, such as the 'Palmatum' group, the leaves are more consistenly five-lobed, while in others, such as the 'Septemlobum' group, the lobes may be mostly seven. The lobes are usually deep and in some forms they may be divided all they way to the base of the blade. The extreme forms, such as 'Dissectum' and 'Sessilifolium,' with strongly dissected lobes, have a delicate, feathery appearance.

In color, the leaves vary from bright green to yellow and different shades of red or purple. They turn yellow to orange or red in the autumn. There are also variegated forms with white spots ('Versicolor'), red blotches ('Bicolor'), pink margins ('Roseo-marginatum'), or various other color combinations. A most becoming form is 'Tricolor,' which has the leaves spotted red, pink, and white.

OTHER ASIATIC SPECIES

Eastern Asia is the center of development of the genus, with about 100, or two-thirds of the total species (Fang, 1939; Matsumura, 1954). A number of these have long been in cultivation in their native countries. In Japan, besides *A. palmatum, A. japonicum* has also been long cultivated. In eastern China, *A. buergerianum* has been in cultivation since early times and was also introduced into Japan centuries ago (Figure 38). The most commonly cultivated species in northern China are *A. truncatum* (Figure 39) and *A. mono.*

In the last two centuries numerous other species growing wild in the mountains of China and Japan have been introduced by plant explorers into cultivation. These include some of the most handsome small and medium-sized trees. Some of these have become highly valued in recent years

38. Acer buergerianum.

39. Acer truncatum.

40. Acer miyabei.

on account of their small sizes and neat appearances, features that are especially desirable for modern landscaping.

From Japan, among the most notable newer introductions are such species as *A. miyabei* (Figure 40), *A. rufinerve,* and *A. tschonoskii. A. cissifolium* and *A. nikoense* have three-foliolate leaves and are related to the American box elder. *A. crataegifolium* and *A. carpinifolium* are distinctive species with unlobed leaves resembling *Crataegus* and *Carpinus* in appearance, respectively.

From Korea, Manchuria, and northern China come such species as *A. ginnala, A. mandshuricum,* and *A. tegmentosum. A. mandshuricum* is a three-foliolate-leaved species. These species, originating from a more northern climate, are among the hardiest in cultivation.

The largest number of new introductions in modern times has come from the mountains of central and western China, the center of development of the maple genus. Among the numerous desirable species, the following are especially notable: *A. oliverianum, A. davidii, A. griseum,* and *A. henryi.* These are all handsome trees mostly of relatively small size. *A. davidii* is a very colorful and distinctive looking tree in the autumn, with its foliage turning yellow and purple. The white, shiny branches are also conspicuous. *A. griseum,* the paperback maple, is equally distinctive in the winter with its flaky bark. *A. henryi* is a three-foliolate-leaved species related to the box elder (Figure 41).

The species from eastern Asia now in cultivation are too numerous to mention. Most of these are still of restricted planting and are found only in arboretums and botanical gardens. Some of the other species which deserve listing are *A. mayrii* (Figure 42), *A. sinense, A. diabolicum, A. longipes,* and *A. capillipes.*

41. *Acer henryi.*

42. *Acer mayrii.*

SPECIES OF THE HIMALAYAN REGION

Species indigenous to the Himalayan region are generally tender and therefore not suitable for the more strictly temperate climates. These often also extend to parts of China and even to Japan, and the forms or races of these latter varieties are hardier than the typical ones from the Himalayan region.

EUROPEAN AND WESTERN ASIATIC SPECIES

The Norway maple, *A. platanoides,* and the sycamore maple, *A. pseudo-platanus,* native to Europe and western Asia, are among the most widely planted trees.

The exact limits of distribution of the sycamore maple (*A. pseudo-platanus*) are difficult to define as the tree has been extensively planted for centuries all over Europe. The native habitat seems to extend from the Pyrenees to the Alps and the Carpathians, and from the hilly districts of France, Germany, and Italy to Greece, the Crimea, and the Caucasus. In other areas in Europe it is extensively cultivated and flourishes as far north as Norway and Sweden (Elwes & Henry, 1906).

There seems to be little variation in the foliage in the wild form, but numerous varieties have arisen in cultivation, especially with respect to the shape of the leaves. A few colored and variegated forms are also known. The best known colored form is 'Purpureum,' with leaves purple beneath. The petioles and fruits are often bright red. This variety originated in 1828 in Sander's Nursery in Jersey, England (Elwes & Henry, 1906).

The Norway maple (*A. platanoides*) (Figure 43) has a wider distribution, covering most of Europe and extending

43. Acer platanoides "Erectum."

44. Acer campestre.

45. *Acer rubrum.*

46. *Acer saccharinum.*

eastward into the Caucasus, Armenia, and northern Persia. It is indigenous in Norway, Sweden, and Finland, and is common in Germany, France, northern Spain, and northern Italy (Elwes & Henry, 1906). In England it is not native but was introduced in 1683 (Loudon, 1838).

The Norway maple is one of the hardiest trees in cultivation and will grow on the driest and poorest soil. It was introduced into North America in 1870 (Rehder, 1940) and is now one of the most widely planted street trees. A number of varieties have arisen in cultivation in Europe including the highly ornamental 'Palmatifidum', the eagle's-claw maple, with its elegantly dissected foliage; 'Schwedleri,' with its colorful leaves, bright-reddish when young, changing eventually to dark green; and 'Erectum,' a narrow-pyramidal tree with large leaves.

The hedge maple, *A. campestre,* of Europe and western Asia, has also been long cultivated, especially in hedgerows (Figure 44). It runs into many forms. Species from the Mediterranean region, such as *A. orientale, A. opalus, A. monspessulanum,* and *A. heldreichii* are generally suitable for the milder climates. Species from the Caucasian region, such as *A. trautvetterii, A. velutinum,* and *A. tataricum* are sometimes also not hardy in the colder regions.

SPECIES OF EASTERN NORTH AMERICA

A few species from eastern North America are now widely planted not only in America but also in Europe and other continents. Among these, the most notable are the box elder, *A. negundo,* and to a lesser extent, the red maple, *A. rubrum* (Figure 45), the silver maple, *A. saccharinum* (Figure 46), and the sugar maple, *A. saccharum* (Figure 47). These species were first cultivated in colonial

47. Acer saccharum.

times in the seventeenth and eighteenth centuries.

The box elder occupies a wide range from eastern to western North America with several geographical varieties and a number of horticultural forms, some of which are now favorite ornamental trees. The species is very hardy and drought-resistant and is much used in the northwestern United States for shelter belts. It was one of the first North American trees introduced into Europe and was cultivated by Bishop Compton in his garden at Fulham near London before 1688 (Loudon, 1838). The variegated-leaved form is now a popular garden plant in most European countries.

The sugar maple, the source of maple sugar, is much planted, especially in the northeastern States, as a street and shade tree. Formerly it was considered as a wide-ranging and variable species, but those forms outside of northeastern United States are now treated as distinct species such as *A. nigrum, A. grandidentatum, A. leucoderme,* etc. The sugar maple was introduced into England by Peter Collinson in 1735 (Aiton, 1789).

The silver maple and red maple are also widely cultivated as ornamental trees and both species have developed a number of horticultural forms in cultivation. The red maple, as it inhabited swamps close to the coast, attracted the attention of early travelers in America. It was carried to England as early as 1658 and planted near London in the garden of the younger Tradescant (Sargent, 1891). It grows as well in Europe as in its native country and many varieties have been found in both American and European nurseries.

Less commonly planted are such species as *A. nigrum* (black maple), *A. leucoderme, A. spicatum* (mountain maple), and *A. pensylvanicum* (moosewood or striped maple).

SPECIES OF WESTERN NORTH AMERICA

Species of the western part of North America are generally suitable for the milder climates only. Species that can barely survive in the Philadelphia region include the Oregon maple, *A. macrophyllum,* and the vine maple, *A. circinatum.* The first is a large tree with very large leaves, attaining to 20 to 30 cm. across in its native habitat (Figure 48). The latter is a small tree with a wide-spreading habit. The

48. Acer macrophyllum. (Photo, Chester E. Pancoast.)

box elder, *A. negundo,* as mentioned before, is the most widely distributed species of maple in North America, extending all the way across the continent. There are also local varieties in California, Texas, etc., but these forms are not hardy in colder regions.

The western American species have been cultivated only since the early part of the nineteenth century. Besides those mentioned above, other species are occasionally also planted, such as *A. grandidentatum,* of the Rocky Mountain area, and *A. glabrum,* which is widely distributed from Alaska to California.

8

The Black Locust and Honey Locust

The early American colonists, like all other immigrating people, were at first more interested in trees of their original homes than in the native plants of their adopted land. Since the very beginning efforts were made to introduce nearly all important cultivated species of European trees into America, in spite of the fact that practically all of them have their counterparts in eastern North America. Moreover species of the same genera that are indigenous are more suitable to the local climate. The American flora is actually far richer than that of Europe. Not only are more genera represented, but there are usually more species in the same genus in America than in Europe.

Only a few of the native American trees were thus cultivated during early colonial times, while most other were not introduced into cultivation until the eighteenth or nineteenth centuries. The cultivation of native trees gained great momentum toward the end of the last century as a result of wider interest in gardening, conservation, and tree culture as well as the establishment of numerous arboretums and botanical gardens throughout the country. Since then nearly every important native species of trees has been in cultivation one way or another.

The trees of American origin therefore have a much shorter history of cultivation than most other important trees from Europe and Asia. Within the short span of about three hundred years, however, two outstanding species have already gained worldwide recognition, and are sometimes more extensively cultivated and better known in some other parts of the world than in their native home. The two species, both plants of the pea or legume family, are the black locust and the honey locust.

THE PEA FAMILY

The pea family, Leguminosae, is one of the largest families of flowering plants. The very numerous genera and species, estimated at over 500 and 8,000, respectively, are widely distributed all over the world and are most abundant in the tropics.

As food plants the many edible legumes are second only to cereals in importance. Because of the fact that they are very rich in proteins, their use supplements the high-carbohydrate content of cereals. In many parts of the world peas and beans actually take the place, in part or wholly, of foods derived from animals.

The Leguminosae have a unique susceptibility to certain root-inhabiting, nodule-forming bacteria that synthesize nitrates from free nitrogen. This process is an extremely important one, because atmospheric nitrogen cannot be used by higher plants in the lengthy process leading to protein formation.

Part of the nitrates formed by the bacteria can be utilized by the leguminous host plants, and part eventually by subsequent crops. Thus, the high nitrogen content of legumes not only contributes to the food value of beans and peas, but also makes these plants useful as fodder and for enriching the soil. The importance of legumes to human economy and agriculture is indeed very great.

The black locust, *Robinia pseudoacacia* (Figure 49), is one of about twenty species of a genus which in North America is distributed from Pennsylvania south to Mexico. Some of the species are of tree size while others remain as shrubs. The black locust is the most widespread and the most commonly planted species. It is also known as locust, common locust, yellow locust, white locust, and acacia. The

49. *Robinia pseudoacacia.*

name "locust" was given by early missionaries who fancied that the tree was the one that supported St. John in the wilderness (Loudon, 1838). It is, of course, an American tree, which is not native to any other part of the world. The true locust of the New Testament is most probably the carob, *Ceratonia siliqua.*

The black locust is a medium-sized tree, usually attaining a height of about 45 feet, but may reach as much as 75 feet. The reddish dark-brown bark, deeply furrowed and ridged, is quite distinct in appearance. The branches and branchlets bear sharp spines at the nodes, usually in pairs. The feathery, odd-pinnately compound leaves, dark green in color, and the large drooping clusters of very fragrant white flowers together make the tree a highly ornamental one.

As noted above, the black locust has the widest range of all species in the genus. The original habitat extended from the mountains of Pennsylvania to Georgia and westward to Iowa, Kansas, and Oklahoma, but it is now widely naturalized in many other parts of North America. Several insects, the locust borer and locust leaf miner, and the fungal locust rot are doing considerable damage to the trees locally. The tree occurs in a wide variety of conditions. It grows especially vigorously on moist fertile soil, along streams, or on rich bottom lands, but it also thrives on rocky, sterile slopes.

USES OF THE BLACK LOCUST

The wood of the black locust is of excellent quality, especially in straight and clean forest-grown specimens free from insects or other enemies. It is very hard and strong, yellowish-brown to reddish-brown, with thin, greenish or yellowish sapwood. It is very durable in contact with soil.

In former times it was used extensively for shipbuilding. It is now used for posts, poles, or ties, for tree nails, insulators, pins, and fuel.

The black locust, like many other legumes, has nodules on the roots. It can be used as a soil improver for pastures or bare lands. The leaves may be used as fodder. The tree is also an important honey plant.

The use of black locust in former times as tree nails in the construction of wooden ships is of particular interest. These nails, the thick, long, wooden pins holding together timbers, were subject to great stress. In colonial times, the black locust nails were used with great success, owing to their great strength together with amazing decay resistance. By 1820 Philadelphia alone exported between 50,000 and 100,000 locust nails to England annually. A member of Parliament rose to point out that unless the British Navy adopted black locust tree nails it could not hope to equal the American ships (Schramm, 1942).

INTRODUCTION OF THE BLACK LOCUST INTO CULTIVATION

The black locust is perhaps one of the very few trees planted by Indians in the temperate regions of North America before the arrival of the colonists. It appears to have been common in the neighborhood of the coast when Virginia was first settled by Europeans (Sargent, 1892). William Strachey, who first visited the colony on the James River in 1610, mentioned this tree in his *Historie of Travaile into Virginia Britannia*. He found that "by the dwelling of the salvages are bay-trees, wild roses and a kynd of low tree, which beares a cod like to the peas, but nothing so big: we take yt to be locust."

The Indians made their bows from the wood of the locust.

According to Sargent (1892), it is quite probable that the Indians of Virginia carried the tree from the mountains to the low country and so helped the tree to spread beyond the limits of its original natural forests along the slopes of the Appalachian Mountains.

The exact date of its introduction into Europe cannot be ascertained. Some authors, such as Linnaeus and Miller, claim it was 1601 when the botanist Jean Robin, after whom the genus was named, is said to have obtained seeds from America; others such as Adanson and Deleuze maintain that it was not until 1636 when it was planted in Paris by Vespasien Robin, son of Jean Robin. This particular tree, the oldest known of its kind in Europe, is still living in the gardens of the Musée d'Histoire Naturelle (Figure 50). The English, however, might have introduced the tree into England at an earlier date as John Parkinson, who published thé first description of the tree in 1640 in his *Theatrum Botanicum,* observed that the locust had been raised near London by Tradescant to already "an exceeding height" (Elwes & Henry, 1912).

In the seventeenth and early eighteenth centuries, the locust had attracted great public attention in Europe and numerous papers had been published on its horticultural value (Elwes & Henry, 1912). In 1892 Sargent was to write that, "No other North American tree has been so generally planted for timber and ornament in the United States and in Europe; and no inhabitant of the American forest has been the subject of so voluminous a literature."

The tree has since been planted in nearly all countries in Europe, and in South Africa, North Africa, South America, Asia, Australia, and New Zealand (Spaulding, 1950). In many countries the black locust is the most extensively used tree for reforestation (Blümke, 1956).

50. The Robin black locust in Paris. (Photo, Edward H. Scanlon.)

THE SHIPMAST LOCUST

A number of geographical strains or races of the black locust have been reported but most of these are probably differences resulting from environment. Among these, the most outstanding and distinct is the variety 'Rectissima,' the shipmast locust, described from Long Island by Raber in 1936 but introduced there apparently over two centuries ago, possibly from tidewater Virginia. The lasting quality of the wood of this tree is probably unsurpassed by any other in the eastern United States. Fence posts over one hundred years old have been resold for further use. This is a wood that can be adapted to many special uses.

VARIATIONS OF THE BLACK LOCUST IN CULTIVATION

A large number of varieties of the black locust have arisen in cultivation (Sargent, 1892; Rehder, 1940). The more important and distinct ones are here enumerated. Nearly all of these originated in European, especially French, gardens in the nineteenth century. They are mostly propagated by grafting.

'Crispa': An unarmed form with undulate or crinkled leaf margins; originated before 1825.

'Decaisneana': A vigorous tree with light rose-colored flowers; originated before 1860.

'Dissecta': A compact tree with short branches and dissected leaves; originated before 1869.

'Inermis': Unarmed; originated before 1904. This is the form usually planted in Europe for fodder.

'Latisiliqua': Pods broad; originated before 1829.

'Macrophylla': Leaves long and leaflets broad; originated before 1830.

'Pendula': With somewhat pendulous branches; originated before 1820.

'Pyramidalis': A form with unarmed erect branches forming a narrow pyramidal head; originated in 1839.

'Tortuosa': A distinct form with short twisted branches, short in stature and slow in growth and usually few-flowered; originated before 1810.

'Umbraculifera,' parasol acacia: An unarmed form with short branches forming a dense spherical head; originated before 1810. This is the form much used in Europe by grafting high for formal planting. This is also one of the most popular street trees in central and northern Europe. It rarely produces flowers (Figures 51-53).

'Unifoliola' ('Monophylla'): A form with leaves sometimes reduced to one large leaflet or more often with two

51. Parasol acacia, Novi Sad, Yugoslavia—old trees. (Photo, Vogin Vasilic.)

52. *Parasol acacia, Novi Sad, Yugoslavia—young trees not trimmed for four to five years. (Photo, Vogin Vasilic.)*

53. *Parasol acacia, Novi Sad, Yugoslavia—trimmed young trees. (Photo, Vogin Vasilic.)*

54. *Robinia psudoacacia 'Unifoliola'—variation in leaves.*

55. *Robinia pseudoacacia 'Unifoliola.'*

or three leaflets or more; originated in 1855 (Figures 54, 55).

OTHER SPECIES OF ROBINIA

With the exception of the black locust, the other species of the genus *Robinia* are mostly shrubs or, at most, small trees. They are cultivated only occasionally, and generally used locally as ornamentals for their showy and often fragrant flowers. The most commonly planted one is the clammy locust, *R. viscosa,* a small tree which grows to a height of about 35 feet. The tree can be readily distinguished from the black locust by its stiff, glandular, viscid, reddish-brown hairs densely covering the young branches. The leaflets are more numerous (13 to 25) than in the black locust (7 to 19). The tree is native to southeastern United States and has been cultivated since the late eighteenth century.

Among the shrubby species, the more commonly planted are *R. hispida* and *R. fertilis,* both natives of the southeastern United States. These species grow to a height of only 3 to 6 feet and are planted mainly for their showy, rose-colored flowers.

THE HONEY LOCUST

The honey locust, *Gleditsia triacanthos,* (Figure 56) is a native of the eastern part of North America, extending from Ontario through Pennsylvania to Florida, westward to Kansas and Texas. It is now widely planted and in many places it has escaped from cultivation and become naturalized. In some parts of the country it is also called sweet

56. Gleditsia triacanthos.

locust, three-horned acacia, thorn tree, honey shuck, or black locust.

The tree is of medium size, attaining usually a height of about 50 to 100 feet, but sometimes it may grow to a height of nearly 150 feet. The trunk is usually short, branching low, but in close stands it may be rather long and clean. The bark is grayish-brown to almost black, usually rough with a few fissures and thick, firm, broad ridges. It is often covered with stout, simple, or three-branched thorns which are frequently also found on the twigs. The long, even-pinnately, once- or sometimes twice-compound leaves are quite ornamental. The greenish flowers, in male and female clusters, are very fragrant. The distinct large, leathery fruit pods measure one to one and a half feet long.

The honey locust is a fast-growing tree. In its original habitat, it develops best in rich soil along moist bottom lands, but it will grow in any fertile soil which is not too wet. It is a light-demanding tree.

USES OF THE HONEY LOCUST

The wood of the forest-grown honey locust is of excellent quality, but in open-grown trees it usually becomes too knotty. It is hard, coarse-grained, strong, heavy, and durable in contact with the soil. It is bright reddish-brown in color with thin, pale sapwood. It is used mainly for fence posts, poles, ties, fuel, and lumber for general construction.

The honey locust is highly ornamental but as a shade tree it is not satisfactory on account of its late leafing in the spring. The tree is an important honey plant. According to Smith (1950), it lacks the root nodules common to most legumes. The pods are edible and can be used as a valuable stock food.

OTHER SPECIES OF GLEDITSIA

The genus *Gleditsia* is composed of about twelve species, distributed in North and South America, and in the Old World in central and western Asia and in tropical Africa. Besides the honey locust, there are two other species in North America. The water locust, *G. aquatica,* inhabits river swamps of southeastern and southern United States. The Texas honey locust, *G. texana,* occurring in the United

States from the Mississippi to eastern Texas, is believed by Sargent (1902) and Elwes and Henry (1912) to be a putative hybrid between the honey locust and the water locust.

A number of species from eastern Asia are now in general cultivation. The most important cultivated species there is *G. sinensis,* a medium-sized tree quite similar to the American black locust in general appearance. This is the famous "soap-pod tree" in China, where the pods have been used since ancient times as soap by boiling in water and producing lather. This is used especially for cleaning furniture as it will not damage any wood structure.

INTRODUCTION OF THE HONEY LOCUST INTO CULTIVATION

The honey locust was cultivated early by American colonists sometime in the late seventeenth century. The first account of it, drawn from the cultivated tree, was published by Plukenet in 1700 in his *Amalthum Botanicum,* with a brief description and illustration.

The tree was first cultivated in Europe by Bishop Compton in his garden at Fulham near London in about 1700 (Aiton, 1789; Loudon, 1838). It was known in France in the early eighteenth century and spread shortly afterwards to other European countries.

Since its first introduction into cultivation, both in America and in Europe, the honey locust has been extensively planted. Its readiness to grow from seed, rapid growth, easy culture, and extreme hardiness are among the commendable characters that make it popular for planting in gardens, parks, or along highways. Its only drawbacks are its lateness in sending out leaves in the spring and a serious insect pest, the mimosa webworm, in the Philadelphia area. An especially desirable character of the tree is its ready adaptibility to any soil conditions.

The honey locust is cultivated in most countries as an ornamental tree. It has been reported to be useful as a fodder in Australia and other places. It is doing well in such wide-ranging areas as the steppes of southern Russia and Tunisia in Africa (Smith, 1950), and is reported also in cultivation in South Africa and South America (Spaulding, 1956).

VARIATIONS OF THE HONEY LOCUST IN CULTIVATION

G. triacanthos is a more or less uniform tree with few variations. In its natural populations or in cultivation, there appear occasional individuals which are unarmed or nearly unarmed, and these trees are usually of slender habit. These plants are referred to as f. *inermis*. In cultivation, only a few variations have appeared as follows:

'Bojutii' ('Pendula'): A tree with graceful pendulous branches and small narrow leaflets; originated before 1845 from cultivated plants in France.

'Elegantissima': Of dense bushy habit, unarmed and with smaller leaflets; originated about 1880 in France.

'Nana': A somewhat smaller more compact shrub or tree with shorter, broader, dark green leaflets; originated before 1838 in England.

9

The Metasequoia

The discovery in 1941 of a living species of *Metasequoia* aroused widespread interest in botanical and horticultural circles. The revelation of the existence of a new type of conifer is in itself highly significant, but the impact of this discovery upon paleobotany, the study of the past history of plants, is even greater. Its very existence means that many long-established concepts of Tertiary botany, based upon fossils alone, must now be re-examined and revised. Thanks to *Metasequoia,* we have gained new insight into the conditions which lead to the development and formation of our present-day vegetation. The story of its discovery has been frequently recounted in many languages, but it is one that deserves to reach an ever-widening public.

The universal interest engendered by this intriguing conifer is reflected in a voluminous and rapidly increasing literature. Since its formal designation as *M. glyptostroboides* in 1948 it has been a constant subject of scientific study and publication. In 1950 Chu and Cooper enumerated 54 articles devoted to it and in 1952 Florin cited no fewer than 130; the number must have increased severalfold since that time.

Within a few years of its discovery, seeds of *Metasequoia* were distributed to all important botanical gardens and arboretums throughout the world. It is now widely cultivated in Europe, Asia, and North America as well as in some parts of the Southern Hemisphere. Since stock is now available from several trade sources its cultivation has also spread to many private gardens. Probably no plant has ever been accorded such rapid and widespread recognition and dissemination, a phenomenon made possible only in this modern age of swift worldwide transportation and communication.

ESTABLISHMENT OF THE GENUS

To begin our story we must go back at least a century. Almost a hundred years ago a fossil plant, later known as *Sequoia langsdorfi* (Brongniart) Heer, made its appearance in the literature. This plant was shown to be widespread in the Eocene at high northern latitudes and at middle latitudes in North America and Asia. Subsequently many other similar fossil plants were discovered and described by paleobotanists. Most of these were named as species of *Sequoia* for they were believed to be related or ancestral to the living redwood *Sequoia* (including *Sequoiadendron,* the big tree) of California. Others were referred to as *Taxodium,* in the belief that they were assigned to the same genus as the living bald cypress of the southeastern United States and Mexico. These fossil plants, on the strength of their presumed identity with these modern genera, were used by paleobotanists as a basis for constructing various theories about Tertiary vegetation in the Northern Hemisphere, its ecology, migration, and evolution.

Many of these fossil plants, including the most widely recorded *S. langsdorfi,* however, do not exactly resemble the modern *Sequoia, Sequoiadendron,* or *Taxodium,* for they bear, among other things, oppositely arranged branches and leaves, while contrastingly, all these modern genera have alternately arranged ones. This very obvious and distinct taxonomic character, interestingly enough, was completely overlooked by paleobotanists for nearly a century. Even though the material available to paleobotanists is always fragmentary, this character nevertheless is readily recognizable in nearly every specimen.

It remained for a taxonomist, the Japanese botanist Miki, a specialist on living aquatic plants, to make the long overdue diagnosis. He by chance discovered and studied some fossils and noted the striking character of opposite leaves of some that had long been credited to *Sequoia*. He also noted that the cones have decussate scales and long slender stalks. As all these characters are manifestly different from *Sequoia* or *Taxodium,* he established a new genus *Metasequoia* (1941), "meta," from the Latin meaning "beyond," in allusion to their possible relationship.

DISCOVERY OF A LIVING SPECIES

When Miki established the genus *Metasequoia,* it was still known only as a fossil genus, one which supposedly had long been extinct. But strangely enough in the very same year, the year of a great war in eastern Asia, on the other side of the battle front, a Chinese botanist was also discovering a living tree, entirely unknown to science, but belonging to this very same genus. It is a strange coincidence, and indeed a fortunate one for science, and apparently one of the few good things coming out of the last, destructive World War.

In 1937 Japan invaded China. In subsequent years the Chinese government retreated to the interior in the mountainous west and dedicated itself to prolonged resistance.

Exploration parties were dispatched to little-known regions in the hinterland of the country to survey and explore natural resources. At the end of 1941, T. Kan of the National Central University, while visiting the village of Mo-tao-chi, in eastern Szechuan province, first noted a peculiar deciduous conifer locally called *"shui-sha"* or water fir; no specimen was then collected and it was not until the summer of 1944 that T. Wang of the Central Bureau of Forest Research collected foliage specimens and cones from the same locality. The discovery of this new conifer was first made known through publication in 1946, and was named and described as *M. glyptostroboides* by H. H. Hu and W. C. Cheng in 1948. The specific epithet referred to its resemblance to *Glyptostrobus,* a deciduous conifer of southern China.

INTRODUCTION INTO CULTIVATION

From 1946 to 1948 Cheng sent out several expeditions to the area to collect herbarium specimens and seeds. In the meantime Dr. E. D. Merrill of the Arnold Arboretum, upon receiving herbarium specimens collected by C. J. Hsieh in 1946, realized the significance of the find and the importance of obtaining seed for propagation. With the aid of a grant made by the Arnold Arboretum, an expedition was sent out by Cheng for obtaining seed. Hsieh and C. T. Hwa were members of the party. Their exploration revealed the existence of a hundred or so large trees of *Metasequoia,* covering an area of about 800 square kilometers. The center of distribution is in the Shui-sha Valley in Hupeh province, where over 1,000 trees were found, many of them of small size and not a few planted.

A large quantity of seed was obtained and subsequently

widely distributed in China, the United States, and many other countries. The wide distribution of seed to botanical gardens throughout the world made by Dr. Merrill is largely responsible for its present great popularity today.

Since then the tree has been successfully raised in many parts of the world. In spite of the small area of its original habitat, it has proved hardy in widely separated regions of different continents, including such northern localities as Norway, Sweden, Finland, southern Alaska, and central Vermont. It likewise thrives in all temperate countries.

Metasequoia is a rapid-growing tree when young. Among the plants grown from seed sown in April, 1948, at the Morris Arboretum, the tallest reached a height of 5 feet 9 inches in 1949 (Skinner, 1949) and in 1957, it measured about 20 feet (Figure 57). A tree planted upon a slope has grown as fast as one planted near a pond. As much as 57 inches of growth has been observed in a single season. This rapid rate of growth is unsurpassed by any other conifers we have noted. The trees are not only rapid-growing but also of well-formed, pyramidal shape. When the leader is damaged, it is rapidly replaced without any deformation.

In many localities trees grown since 1948 have already been reported to bear female cones. At least three such instances have come to our notice, one along the Pacific Coast, a second in New York, and a third in England. In none of these cases were male flowers noted, so the cones were apparently sterile.

Although it may be many more years before most of the trees in cultivation will bear male as well as female cones and thus produce viable seeds, the tree fortunately is readily propagated by cuttings so there is no shortage of propagating stock. Green-wood cuttings taken in late summer readily took root after auxin treatment, and vigorous young trees were obtained in a year or two. Cuttings taken from term-

57. *Metasequoia at the Morris Arboretum. (Photo, J. R. Carpenter.)*

inal as well as lateral branches are reported to develop equally into erect well-shaped trees.

In cultivation the trees seem to be free from any fungus disease. Few insects seem to attack them and although the ubiquitous Japanese beetle has been reported to do damage to the trees, we have not observed this adverse fact locally. Altogether, its shapely form, its bright green feathery foliage turning brownish-yellow in the fall, its ease and rapidity of growth, and its freedom from disease and pests have already made this a popular ornamental tree. It may also prove to be of value as a timber or pulp tree. In any case *Metasequoia* seems to have become firmly established as one of the cultivated trees of mankind.

ORIGINAL RANGE AND HABITAT

In order successfully to grow a plant in a wide variety of localities and situations one requires first of all a knowledge of its original range and habitat. In the case of *Metasequoia,* there are available several papers on the subject based on first-hand observations.

Early in 1948, Chaney made a hurried trip to the homeland of *Metasequoia,* spending a few days in the field. Since it was winter, little could be observed or collected (Chaney, 1948). In the summer of the same year, Cheng, together with Hwa and Chu, explored the area for several months and later Chu reported his observations in a joint paper with Cooper entitled "An ecological reconnaissance in the native home of *Metasequoia glyptostroboides"* (1950). This is the most detailed study of its kind. During the same summer, Gressitt also conducted an expedition to the region and spent considerable time in the field collecting primarily insect specimens. His travels and observations are reported in a paper published in 1953.

The first map showing the range of *Metasequoia* is given by Hu and Cheng (1948) in the paper which contains the original description of *M. glyptostroboides.* Chu and Cooper (1950) later gave a more detailed map showing its range. While Gressitt's paper does not contain a map, his account of his journey shows that the area covered by him is more extensive than Chu's. He actually discovered and observed many scattered specimens outside the range given by Chu. By carefully checking over Gressitt's itenerary and observations, it is found that the range needs extension southward beyond Lat. 31°, south of the city of Chung-lu. A revised map based on the original ones given by Hu and Cheng and Chu and Cooper and incorporating Gressitt's additional observations is given here (Figures 58, 59). The

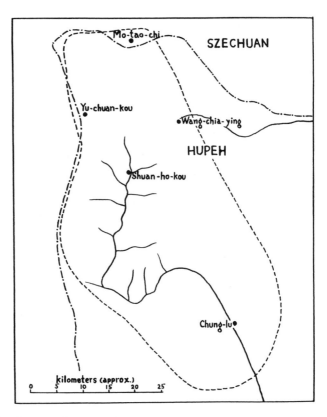

58. A map showing the approximate limits of Metasequoia.

area of the original range of *Metasequoia* in this map is about one-third larger than that given by Chu and Cooper, estimated by them as covering 800 square kilometers.

Metasequoia is found in west-central China in the north-western corner of Hupeh province, ranging at points slightly across the border of Szechuan province. It occurs primarily in the Shui-sha Valley, a narrow, close depression at a relatively high elevation (2150 to 4250 feet) in a sandstone region. The natural habitat of the tree is in shady, moist localities on slightly acid or circumneutral

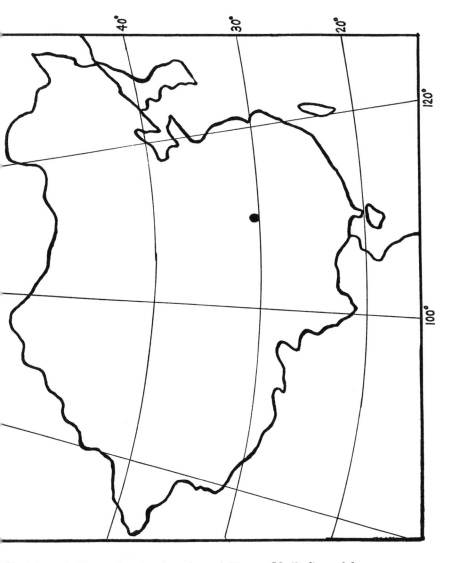

59. *Map of China showing location of* **Figure 58** *(indicated by black dot.)*

soils. In this region the species occurs as a constituent of a presumably natural and actively reproducing forest community (Figures 60-66).

The tree grows mainly in ravines, from high up the mountain slopes down to the floor of the valley. It thrives best along the rocky banks of small streams and seepage grounds at the foot of the slopes forming dense thickets with other trees and shrubs. It is essentially a member of a stream-bank community (Figures 63-65). Moist soil and humid atmosphere seem to be essential for its successful reproduction. Seedlings are found in such moist localities. Gressitt reports that though the trees are generally erect, sometimes young individuals have widely spreading branches making them as broad as they are high, like yew trees in the valley.

The trees frequently associated with *Metasequoia* in-

60. A view of the Shui-sha Valley. Note the terraced rice paddies. (Photo, J. L. Gressitt.)

61. One of the largest Metasequoia trees, located in lower Shui-sha Valley.

62. Another view of the Shui-sha Valley.

63. A grove of Metasequoia trees. (Photo, J. L. Gressitt.)

64. A grove of Metasequoia trees.

65. *A Metasequoia tree planted near farm house by the side of rice fields. Note how the tree is closely cropped. (Photo, J. L. Gressitt.)*

66. *High mountain pass between ridges leading to Shui-sha Valley.*

clude three conifers, *Cunninghamia, Cephalotaxus fortunii,* and *Taxus,* and some thirty angiosperms including *Liquidambar formosana, Castanea sequinii, Rhus* spp., *Cornus controversa,* and *Lindera glauca.* The commonest shrubs are species of *Spiraea, Hydrangea, Viburnum, Morus, Berberis,* and many others. Woody climbers include *Hedera, Rosa, Parthenocissus, Akebia, Actinidia, Lonicera,* etc.

It can thus be seen that the natural vegetation is primarily moist mesophytic and warm temperate in nature. No climatic data are available from the region, but observations of and comparisons with nearby areas show that it is a region of mild winters and hot summers, with the highest precipitation in spring and summer months. Frosts are probably infrequent and light. Chu and Cooper believe that a near climatic parallel in North America is the southeastern coastal plain, e.g., Georgia.

CULTIVATION IN ITS ORIGINAL RANGE

In its original center of distribution, the Shui-sha Valley, there has been considerable planting of *Metasequoia.* Seedlings and young trees are constantly being removed from native sites to be planted along the banks of the main stream in straight rows and along the margins of the rice paddies, along roadsides, and in homeyards. The tree, however, is of little economic importance there. It is planted for aesthetic purposes and is used only incidentally for fuel and in construction. The wood is not considered to be of good quality.

There is a prevalent local custom of periodically cutting the branches of *Metasequoia,* as well as *Cunninghamia,* often almost to the top. Most of the trees outside the shady ravines have thus an extremely slender appearance, a trunk with new branches growing out laterally (Figure 66).

PROSPECT IN CULTIVATION

The rapid growth of popularity in the cultivation of *Metasequoia,* aside from deliberate dissemination and effective advertising, is due to its novelty and its intrinsic ornamental value. Because of the former it will always hold an indispensable position in any botanical garden, and because of the latter, it is now fast becoming a most desirable tree for public parks and private gardens all over the world. The ease of cultivation, rapid growth, and freedom from pests and diseases undoubtedly greatly enhance its value as an ornamental tree.

In addition to becoming a popular specimen tree in botanical gardens and on private lawns, *Metasequoia* seems to have further values as a useful cultivated tree. Being a deciduous conifer and generally of a narrow pyramidal shape, it can be used for lining avenues and walks. The lower branches can be trimmed off when planted along city streets or where space is at a premium. It may prove to be of comparable value to *Ginkgo,* now so extensively used for this purpose. The deciduous habit will enable it to withstand city atmospheric conditions which are usually injurious to evergreen conifers.

As the wood of *Metasequoia* is not considered of good quality, the tree is apparently not promising as a timber species, but its rapid growth seems to indicate that it is potentially a desirable pulp tree for certain regions. Its actual rate of growth for a given area can of course only be determined by experimental studies, but its rapid growth as reported in many regions seems to indicate that it can be widely explored for this purpose. Another desirable attribute of *Metasequoia* is its monotypic nature, that is, it is the only species of a genus and has no close relatives. Since it is essentially free from insects pests or fungus diseases its introduction into a new region is therefore not likely to result in the introduction of new diseases which would constitute a threat to the native trees.

10

Conifers

In the Northern Hemisphere the coming of winter, with the shedding of leaves of most trees, presents a desolate scene. Most conifers and some broad-leaved evergreen trees, however, maintain their foliage, bringing cheerfulness and warding off dreariness. It is but natural conifers are used everywhere for winter decorations. From very ancient times, these trees have become symbols of hopes and memories and mystical qualities have been attributed them.

Ancient men perhaps had frequently saved themselves by seeking shelter under these evergreen forest trees in winter blizzards. The resinous branches undoubtedly furnished the best material as torches for the early cave dwellers. Early men learned to admire the immense size and long life of these forest giants. Undoubtedly some such awesome specimens were singled out as the first shrines of worship. The seemingly everlasting quality of these evergreen trees must also have made them the first trees selected to be planted as memorial trees for temples, shrines, and graveyards.

At present conifers are among the most important timber trees of the temperate and colder countries. In the Orient, in China and Japan, conifers are more commonly planted for shade and ornament than in other countries. Avenues of evergreen conifers are commonly seen in parks, temple yards, mausoleums, palace grounds, and other similar locations. All over the world, they are extensively used for ornamental purposes. However, they are usually seldom used as street trees in urban areas, as evergreens do not thrive in the sooty atmosphere of modern cities.

There are, however, a few genera of deciduous conifers. These can be used for city planting when conditions are not favorable to evergreens. The deciduous conifers include *Larix* (larch), widespread in the Northern Hemisphere, *Taxodium* (bald cypress) of North America, *Pseudolarix* (golden larch) (Figure 67), *Glyptostrobus* (Chinese water pine), and the newly discovered *Metasequoia* of China.

EUROPEAN SPECIES

The classic, or Italian, cypress, *Cupressus sempervirens*, of Europe is now well known as the tree that gives distinction to Italian gardens. Native from southeastern Europe to northern Persia, it has been cultivated since ancient times and was carried from Greece to Italy by the Romans. The wood is fragrant and very durable, and mummy cases are believed to have been made from it in ancient Egypt. It is one of the trees noted for long life, with an age limit estimated at 3,000 years. In medieval Persia, the tree was used to symbolize fire, for the shape was thought to resemble a flame, and it was planted near the temples of the fire worshippers. For centuries, in western Asia as well as in Europe, it has been used as the symbol of mourning, and is still planted in the burial places. It is mentioned more frequently in the classical literature than any other conifer (Loudon, 1838, 1844).

The common yew, *Taxus baccata*, is another long-lived tree and it was regarded in some parts of Europe in ancient

67. Pseudolarix amabilis. (Photo, J. R. Carpenter.)

times as the emblem of longevity. The name "yew" is believed to come from the same root as *"ewig,"* the German word meaning "everlasting." Yews were planted in the churchyards, especially in the south of England. In the English folklore, the yew was the saddest of all trees except the cypress, and was used along with burial ceremonies. The tough and elastic wood made bows used in early wars.

The common yew is now the most popular hedge plant in many places. Many garden forms have developed in cultivation, varying from dwarf or low-spreading forms to erect, fastigiate ones. Variations in foliage size and color are also very numerous.

In Europe, other species of conifers of long cultivation include the silver fir, *Abies alba,* of the mountains of central and southern Europe; the Norway spruce, *Picea abies,* of northern and central Europe; the Scotch pine, *Pinus sylvestris* (Figure 68), of most parts of Europe and also Siberia; and the European larch, *Larix decidua,* of northern and central Europe. The genus *Larix,* as mentioned above, is one of the few genera of deciduous conifers.

These species have all been cultivated since very early times and there are now numerous garden forms originated in cultivation in each of these species. They are also widely planted in other parts of the world. Many other species of European origin are also frequently planted in gardens, but their cultivation is not as ancient as those mentioned.

Nearly all European species of conifers are now in cultivation, not only locally but also in other continents, especially in North America, where many species are widely and extensively planted. The following is a nearly complete listing of European species:

Abies alba	Silver fir
A. cephalonica	Greek fir
A. pinsapo	Spanish fir

68. Pinus sylvestris.

Cupressus sempervirens	Italian cypress
Juniperus communis	Common juniper
J. excelsa	
J. macrocarpa	Plum juniper
J. oxycedrus	Prickly juniper
J. phoenicea	Phoenician juniper
J. thurifera	
Larix decidua	European larch
Picea abies	Norway spruce
Pinus halepensis	Aleppo pine

P. heldreichii	
P. mugo	Mountain pine
P. nigra	Austrian pine
P. peuce	Macedonian pine
P. pinaster	Cluster pine
P. sylvestris	Scotch pine
Taxus baccata	English yew

NORTH AFRICAN SPECIES

North Africa is not rich in conifers, but there is one species that has become widely known in recent times as an ornamental tree all over the world. This is the Atlas cedar, *Cedrus atlantica*. It occurs in the mountains of Algeria and Morocco at high altitudes as isolated groves or forests. The exact date of its introduction into cultivation is not known, but it is most probably around 1844, into England (Elwes & Henry, 1906). Among other North African conifers that are sometimes planted are the Algerian fir, *A. numidica*, first introduced into cultivation in 1862, and *J. thurifera*, which also occurs in southwestern Europe.

WESTERN ASIATIC SPECIES

There are a number of western Asiatic conifers now in cultivation in temperate gardens around the world. Some of these are of wide ranges, extending also to parts of Europe, such as *J. oxycedrus* (prickly juniper), *J. excelsa*, and *J. sabina*. Others, more limited in distribution, are confined to western and central Asia, such as *A. cilicica* (Cilician fir), *A. nordmanniana* (Nordmann fir), *P. orientalis* (Oriental spruce) (Figure 69), *P. schrenkiana*, and *J. dru-*

69. Picea orientalis.

pacea (Syrian juniper). These have mostly been introduced into European and American gardens in the nineteenth century. The Nordmann fir and Oriental spruce are among the most handsome and popular conifers in cultivation.

The long-cultivated classic cypress discussed above, now widely cultivated in the Mediterranean region, is actually unknown in the wild. It has been suggested that it may have been originated in western Asia. The most famous conifer of western Asiatic origin, however, is the cedar of Lebanon. This tree, *Cedrus libani* (Figure 70), a beautiful tree with distinct appearance, is now regarded in many parts of the world as a most esteemed ornamental tree. A native of Asia Minor and Syria, it is well known for its scriptural and historical origins. It is undoubtedly one of the earliest

70. Cedrus libani. (Photo, Alan W. Richards.)

trees planted by man. Solomon transplanted cedars in the low plains of Palestine from the mountains of Lebanon. At that time the tree formed extensive forests. However, through excessive lumbering in the ensuing 3,000 years, it now has become a very rare tree in its native habitat.

Since ancient times, the tree was not only held in high esteem for its beauty and age, but also for the fragrance and lasting quality of the wood. The wood was preferred above all others for building purposes. The magnificent Temple of Solomon was constructed of this wood, as were many other ancient palaces. The planks for the great fleet maintained at Tyre were made of this wood, as well as Solomon's chariots (Moldenke, 1952).

HIMALAYAN SPECIES

Of the scores of conifer species growing in the Himalayan region, one, the deodar cedar, *C. deodara,* has become a most popular and desirable ornamental trees in all temperate regions of the world, and is considered in many countries as the choicest tree in their parks and gardens. The tree was first introduced into cultivation in England in 1831 (Elwes & Henry, 1906). Another frequently and widely cultivated species is the Himalayan pine, *Pinus griffithii* (Figure 71), native to temperate Himalaya from its western part to Afghanistan. The tree is known as "blue pine" in India. It was first introduced into cultivation by Lambert into England in 1823. Other species of Himalayan origin sometimes planted in European and American gardens include the Himalayan fir, *A. spectabilis;* pindrow fir, *A. pindrow*; Himalayan cypress, *C. torulosa*; and Himalayan spruce, *P. smithiana.* All these trees were first introduced into Western gardens in the early part of the nineteenth

71. Pinus griffithii.

century. Most of them are tender and thrive only in relatively mild climates.

CHINESE SPECIES

China is especially rich in coniferous plants. Many species have been in cultivation there since time immemorial, but even to this day, new species and even new genera are still being discovered in the more remote mountainous regions.

Genera endemic to China only include three deciduous ones, *Pseudolarix, Glyptostrobus,* and *Metásequoia,* the last named being a recent discovery, and the evergreens *Biota Cunninghamia, Keteleeria,* and *Taiwania.*

Among the earliest cultivated species in China, as recorded in the classics 3,000 or more years ago, are the arborvitae, *B. orientalis;* the juniper, *J. chinensis* (Figure 72); and the pine, *P. tabulaeformis.* From the very earliest these conifers have been planted as memorial trees for

72. *Juniperus chinensis.*

temple grounds and graveyards. These as well as other species are also used extensively as ornamental trees for the garden. For forestation, the most widely used tree is *Cunninghamia*. Other trees used extensively for this same purpose are *P. tabulaeformis* (Figure 73), *P. massoniana* (Figure 74), and *Cryptomeria japonica* (Figure 75).

Chinese conifers were first introduced into Western gardens in the eighteenth century. Seeds of *B. orientalis* (*Thuja orientalis*) were sent to France from Peking by Pierre D'Incarville around 1752. George Leonard Staunton, who was with the Macartney Embassy sent to China in 1792, introduced into England, among other things, the more tender

73. *Pinus tabulæformis.*

74. *Pinus massoniana.*

75. *Crypotemeria japonica.*

76. *Pinus bungeana.*

conifer *C. funebris,* a graceful weeping tree frequently planted around graveyards in southern China. Robert Fortune introduced the same tree again in 1846-1848, who also introduced the golden larch, *P. amabilis,* in 1853. Among his other introductions to the English garden is the famed lacebark pine, *P. bungeana* (Figure 76), in the same year, but this tree was noted earlier by Bunge in 1831.

Toward the end of the nineteenth century and in the

early part of the twentieth century, numerous new species of conifers were discovered in the borderland provinces of western and northwestern China by successive generations of botanical explorers. A large number of these are now in cultivation in the West (Bretschneider, 1898), but mostly only in botanical gardens and parks.

The more generally planted native conifers and taxads in China are listed below. Nearly all of these are now also widely cultivated over other parts of the world, especially in temperate countries. This list does not include the numerous new species discovered in recent times as mentioned above.

Biota orientalis	Chinese arborvitae
Cephalotaxus fortunii	Chinese plum yew
Cryptomeria japonica sinensis	Chinese cryptomeria
Cunninghamia sinensis	Chinese fir
Cupressus funebris	Chinese weeping cypress
Juniperus chinensis	Chinese juniper
J. formosana	Formosan juniper
Keteleeria davidiana	Keteleeria
Larix gmelini	Dahurian larch
L. potaninii	Chinese larch
Pinus armandii	Armand pine
P. bungeana	Lacebark pine
P. massoniana	Masson pine
P. tabulaeformis	Chinese pine
Pseudolarix amabilis	Golden larch
Torreya grandis	Chinese torreya

JAPANESE SPECIES

The Japanese islands are also exceedingly rich in coniferous plants and there are two endemic genera known only to these islands: *Sciadopitys,* the umbrella pine, and *Thujopsis,* the Hiba arborvitae. Both of these occur naturally only in parts of the center of Honshu, the main island of Japan.

Tree planting has been practiced in Japan from time immemorial. Planted avenues and groves of *Cryptomeria* and *Pinus thunbergii* are common features of Japanese scenery, and old massive specimens of these are often impressive features of many sacred and scenic places in the country. In feudal times in ancient Japan, five beautiful and majestic trees were planted and selected as the most valuable. Known as the "Five Trees of Kiso," they are *Chamaecyparis obtusa, C. pisifera, Sciadopitys verticillata, Thuja standishii,* and *Thujopsis dolabrata.*

The most extensively planted forest trees today are *C. japonica. C. obtusa* (Figure 77), and *Larix leptolepis.* For covering bare hills, *P. densiflora* (Figure 78) is widely grown and *P. thunbergii* (Figure 79) is used especially by the sea. These and many other species are also much cultivated for ornamental purposes.

Most of these common Japanese conifers are now widely planted all over the world for landscaping uses. The introduction of Japanese conifers into the Western garden began in the nineteenth century. Von Siebold first introduced a number into Europe from 1830 to 1855, including such as *Cephalotaxus drupacea, P. densiflora, P. thunbergii, Torreya nucifera,* and *Tsuga sieboldii.* Robert Fortune introduced *Taxus cuspidata* and *Thuja standishii* into England between 1854 and 1860. Many of these Japanese conifers were subsequently introduced into the American garden from Europe, but Dr. George R. Hall in 1862 introduced a number directly from Japan to North America. Among these are *A. firma, Chamaecyparis obtusa, C. pisifera, S. verticillata* and *T. cuspidata* (Wilson, 1916).

77. *Chamæcyparis obtusa.*

78. *Pinus densiflora.*

79. *Pinus thunbergii.*

NORTH AMERICAN SPECIES

The conifers of North America are especially well developed along both the eastern and western parts of the continent, particularly in the coastal states. Some species, however, of more northern ranges, are distributed more widely and continuously from Alaska in the west to the eastern seaboard. The common juniper, *J. communis,* is perhaps the most widespread of all conifers in natural habitat as it grows all over the northern parts of North America and Europe as well as Asia. The species is a very variable one and several garden forms have been derived from various geographical varieties.

Among the northern species with east to west ranges now under general cultivation are black spruce, *P. mariana;* white spruce, *P. glauca;* and American larch, *L. laricina.* These species were first brought into cultivation in the early eighteenth century from trees of the eastern ranges, in northeastern North America, rather than from trees of the western regions.

The eastern species, because of their presence in areas of early settlement, were necessarily brought under cultivation before the western ones. Perhaps the earliest and the most widely planted species is the American arborvitae, *T. occidentalis,* cultivated since 1536. Most of the other eastern species were first planted from the late seventeenth to the eighteenth centuries, a number of them first in Europe rather than in America. The more commonly planted species are listed below. Garden varieties have appeared in most of these species and sometimes these were originated under cultivation in Europe also.

Abies balsamea	Balsam fir
Chamaecyparis thyoides	White cedar
Juniperus virginiana	Red cedar
Picea rubens	Red spruce
Pinus banksiana	Jack pine
P. palustris	Longleaf pine
P. resinosa	Red pine
P. rigida	Pitch pine
P. strobus	White pine
P. taeda	Loblolly pine
Taxodium distichum	Bald cypress
Thuja occidentalis	American arborvitae
Tsuga canadensis	Common hemlock

The indigenous conifer flora of western North America is even richer than in the east. The natural forests of giant redwood, big tree, Douglas fir, and others are among the most magnificent of the world. These species were not in cultivation until the nineteenth century. *Sequoia sempervirens,* for instance, was first discovered by Archibold Menzies and was introduced into English gardens in 1846 by Karl Thoeodor Hartweg. The first authentic account of *Sequoiadendron* was given by William Lobb in 1854, and he also succeeded in introducing it into the English garden. Not only were many of the western North American species first introduced into cultivation in European instead of American gardens, but they also generally thrive better in Europe than in the eastern part of North America. Many horticultural varieties of these American species have since been developed in European gardens. Other species, such as the Monterey pine, have now become most important timber trees in Australia and other parts of the Southern Hemisphere.

The following is a list of the more important species of conifers now in general cultivation originated from western North America:

Abies concolor	Colorado fir
A. grandis	Giant fir
A. lasiocarpa	Rocky Mountain fir
A. magnifica	Red fir
A. nobilis	Noble fir
Calocedrus decurrens	Incense cedar
Chamaecyparis lawsoniana	Lawson cypress
C. nootkatensis	Nootka cypress
Cupressus macrocarpa	Monterey cypress
Juniperus pachyphloea	Alligator juniper
J. scopulorum	Western red cedar
Larix occidentalis	Western larch
Picea engelmannii	Engelmann spruce

P. pungens	Colorado spruce
Pinus cembroides	Nut pine
P. contorta	Shore pine
P. flexilis	Limber pine
P. monticola	Mountain white pine
P. ponderosa	Western yellow pine
P. radiata	Monterey pine
P. sabiniana	Digger pine
Pseudotsuga menziesii	Douglas fir
Sequoia sempervirens	Redwood
Sequoiadendron giganteum	Big tree
Thuja plicata	Giant arborvitae

MEXICAN SPECIES

In Mexico, the coniferous flora is also very rich, but as the species are mostly adapted to warmer climates, they are seldom seen in cultivation in most temperate countries. Among the trees in Mexico, perhaps the most famous of all is the Mexican bald cypress, *Taxodium mucronatum,* especially the few trees that have attained great age and size. The largest specimen is the one at Tule, believed to be two thousand years old. The Cypress of Montezuma, which is the largest of the cypress trees in the gardens of Chapultepec, has been estimated to be over seven centuries old.

Another well-known species of Mexican origin is the Mexican cypress, *C. lusitanica.* This tree has long been cultivated in Portugal and Spain. It forms forests in Bussaco, in Portugal, long celebrated as the home of this tree. However, botanists now generally agree that it is really of Mexican origin. The European tree was introduced, though unknown by whom and when, over three centuries ago from Mexico (Elwes & Henry, 1906).

11

Ornamental Flowering Trees

Most trees have small, insignificant flowers and their use for shade or ornament is mainly because of their foliage effect. Some trees, although primarily effective because of their foliage, may also have relatively showy flowers. Trees like *Sophora* or *Robinia* have large clusters of small but quite conspicuous flowers. Species of *Aesculus* have also large flower clusters, with the individual flowers varying from small, inconspicuous green ones to large, showy red ones. In *Gleditsia* and *Gymnocladus,* the clustered flowers are more or less greenish and not showy. These trees have been discussed under their respective chapters.

Among the other trees, a few also bear relatively large flowers or flower clusters, but because of either inconspicuous coloration, short duration, or the presence of more decorative foliage, their flowers are not valued as ornamentals. *Liriodendron,* for instance, has relatively large single flowers, but the greenish petals and brief appearance render these insignificant.

Many other woody plants, though adorned with brightly colored ornamental flowers, are often of too small a size to be classified as trees. In cultivation, these small tree species especially tend to be shrubby in appearance. Among these belong certain species of the genera *Chionanthus, Cornus, Lagerstroemia, Styrax, Syringa, Viburnum, Xanthoceras,* and many others.

The kinds of trees that grow to a large size and are grown primarily for their ornamental flowers are thus not many. Among these are species which flower in early spring before the appearance of the leaves, such as *Magnolia* and *Paulownia*. Others may flower later in the season with the foliage already on, such as *Albizzia, Catalpa, Davidia, Halesia,* and *Koelreuteria*. Occasionally there may be evergreen trees with distinct flowers, such as *Magnolia grandiflora*.

ALBIZZIA

Albizzia is a tropical genus related to *Acacia,* but although it is primarily of the warm climate, there are two species from temperate China that are widely planted not only in tropical and subtropical countries all over the world but also in various warmer temperate regions. The two species are *A. julibrissin* and *A. kalkora*. The later is relatively rare in cultivation but the former is very commonly and widely planted. In this species, there is also a hardier variety, 'Rosea,' that can be grown in eastern United States as far north as Massachusetts.

A. julibrissin (Figure 80) is generally known as the silk tree because of the showy clusters of pink, silken flowers. The "silk" is the numerous long, slender filaments of the stamens. The large, finely divided feathery leaves are also highly ornamental. Though generally of small stature in cultivation, especially in more northern climates, it may grow to a height of 40 feet or much more in warmer regions. The long flowering season, lasting often up to three months throughout the summertime, is most desirable.

A. julibrissin has a native range extending from central China to Persia. It has been in cultivation in Asia since

80. Albizzia julibrissin. (Photo, J. R. Carpenter.)

ancient times. It was first introduced by D'Incarville into European gardens in 1745.

CATALPA

There are about ten species in the genus *Catalpa,* distributed disjunctly and evenly in two groups, one in eastern North America and one in eastern Asia. The Asiatic species are especially commonly cultivated for shade and ornament, for the street as well as for the lawn. The catalpas are rapid-growing trees with large bell-shaped flowers in panicles. The long, slender, podlike capsules are very characteristic.

Catalpas are among the oldest known cultivated trees in China. The trees were planted in ancient times for their high-quality, easily worked timber. Together with the mulberry, used for raising silkworms and also for timber, the two are considered as the most valuable trees and are universally planted around the Chinese homesteads (Li, 1959).

At present, both in China and in other temperate countries, two of the most widely planted species of the genus are *C. ovata* and *C. bungei.* These trees were introduced into Japan by Buddhist priests in the middle ages and into European gardens in the nineteenth century. The flowers of *C. ovata* are creamy white and marked with orange bands and purple spots. The flowers of *C. bungei* are white with purple spots on the throat. *Catalpa ovata* has fragrant flowers, and the leaves are without the disagreeable color common to the rest of the genus.

The common American species, *C. bignonioides,* is native to southern United States, from Georgia to Florida and Mississippi, but is naturalized as far north as New York. It it a handsome tree, with showy blooms and has been in cultivation since the early eighteenth century. The flowers, slightly smaller than those of the two Chinese species, are white and have two yellow stripes and many purple-brown spots inside.

DAVIDIA

Davidia, or dove tree, is one of the most famous trees introduced into Western gardens from the mountain fastnesses of China in recent times. A rare and curious tree, its introduction and subsequent flowering in cultivation caused great excitement in the horticultural world.

Davidia is a medium-sized tree with leaves resembling those of lindens. The inflorescence consists of a compact head of numerous small male flowers and one perfect flower subtended together by two large unequal bracts up to 6.5

inches long, which are greenish at first and become snowy white at their prime. These pure white bracts resemble large white wings, resting or fluttering among the branches in the breeze—hence the name dove tree.

The species *D. involucrata* (Figures 81,82) was first discovered in 1869 in western China. There is a variety *vilmoriniana,* which is more hardy and was first introduced into France by Père Farges toward the end of the last century and later on also by E. H. Wilson to England and America in 1899-1901 and again in 1903-1904. It is now widely planted in both Europe and North America.

81. Davidia involucrata.

82. Flowers of Davidia.

HALESIA

The several genera of the storax family, Styracaceae, all bear highly ornamental flowers. Most of these, such as the familiar *Styrax* and the less familiar *Pterostyrax,* are shrubs or at most small-size trees. Only in the genus *Halesia* are there species that truly attain tree size.

Halesia, the silver-bell tree, is a genus containing three or four species in North America and one little-known species in China, which is not in cultivation. The genus therefore can be considered as a distinct American contribution to horticulture. Among the species, *H. carolina* and *H. diptera* may grow into small trees only, but *H. monticola* can attain a height of 90 feet. When the tree is laden with tens of thousands of hanging, white, bell-shaped flowers, it is truly a beautiful sight. This tree certainly deserves more general appreciation than it now receives.

H. monticola is native to the southeastern United States

from North Carolina and Tennessee to Georgia. It is a late introduction to the garden and did not come into cultivation until toward the end of the last century. There is a pink form 'Rosea' with pale, rose-colored flowers.

KOELREUTERIA

Koelreuteria is a genus with about five or six species, long considered to be endemic to eastern Asia, but recently a species of this same genus has also been reported to occur in Fiji. The commonly cultivated species is *K. paniculata,* native to a wide area in northern, western, and eastern China. It is one of the earliest planted trees in China, known since the most ancient times as *"luan."* It was designated in the early dynasties some three thousand years ago as one of the five official memorial trees, the one that was once especially planted on the tombs of scholars. The flowers are used for making a yellow dye.

This tree was first introduced into Europe in 1763 by Pierre D'Incarville (Bretschneider, 1898). It is generally called "China tree" or sometimes "goldenrain tree," but it is also known as "pride of India" or "varnish tree."

MAGNOLIA

Magnolias are undoubtedly the most gorgeous of the early, spring-flowering trees. In the size of the flowers alone, no other tree in the temperate regions can compare with it.

The thiry-odd species of the genus occur naturally in two widely disjunct ranges, in eastern North America and southeastern Asia. Most of the American species are evergreen and therefore blossom with the leaves on. The species from Asia are mostly deciduous, and these often produce flowers in early spring on bare branches before the leaves appear. The American species therefore are not as showy as their Asiatic counterparts and are seldom used for their flowering effect. The only exception is *Magnolia grandiflora,* which is valued for both its ornamental evergreen foliage and its large conspicuous flowers.

Among the Asiatic species, two, *M. denudata* and *M. liliflora,* native to central China, are widely planted in China and Japan as well as in other countries all over the world. These two species have been in cultivation in China since the seventh to the tenth centuries (Li, 1959). The first, a larger tree with white, larger, and more showy flowers, is especially valued as an ornamental tree. It is regarded as an emblem of purity and is much used as an art motif, especially in painting and as decoration. *M. denudata* was first introduced to the West in England by Sir Joseph Banks in 1780, and *M. liliflora* in 1790 (Bretschneider, 1898; Johnstone, 1955). From there the cultivation spread to France and other European countries.

The cultivation of these two species in Europe resulted in the production of a chance hybrid *M. × soulangeana.* This was first discovered by Chevalier Soulange Bodin in his garden at Fromont, near Paris in 1820 (Dandy, 1950). Many forms have since been raised and this tree is now one of the most popular spring-flowering trees in both European and American gardens.

An American species, *M. grandiflora,* bull bay, is a magnificent evergreen tree with large lustrous leaves and very large fragrant flowers. It is native to the southeastern United States from North Carolina southward to Florida and Texas. In cultivation since 1734, it is now much planted in the warmer regions of this country as well as in Europe and Asia. It is perhaps the most valued flowering tree of American origin in the horticultural world abroad.

PAULOWNIA

There are about ten species of *Paulownia* in eastern Asia. The most widely and commonly cultivated species is *P. tomentosa* (Figure 83), a stately tree with handsome,

83. Paulownia tomentosa.

large, violet flowers in the spring which appear before the leaves. The species is native to China but is cultivated extensively in Japan also, having been introduced there at an early date. This tree is the celebrated *tung* tree of the Chinese classics, valued since ancient times as an ornamental tree and for its high quality timber. It was introduced into Europe in the mid-nineteenth century and is now widely planted around the world. In eastern North America it frequently escapes from cultivation and becomes naturalized in the woods.

The native range of *P. tomentosa* is central and western China. In eastern and southeastern China, *P. fortunei* is native and is also frequently cultivated. In recent times other species of the genus have also been introduced into cultivation.

PRUNUS

In several of the large genera of the rose family, Rosaceae, such as *Crataegus* (hawthorn), *Malus* (apple, crabapple), *Prunus* (almond, peach, plum, etc.), and *Pyrus* (pear), there are found many of our most familiar fruit trees as well as flowering trees. Most of these, however, are plants of small stature, either of shrub size or not growing over 30 to 40 feet in height. Among those that attain considerable height are several Japanese species of flowering cherries, especially *P. serrulata* and *P. yedoensis*. These often grow to 50 to 80 feet in height.

The most important species, from which nearly all the numerous varieties of Japanese flowering cherries are derived, is *P. serrulata*. The species was first described by Lindley in 1828 from a garden form with double white flowers (now known as 'Albo-plena'), but later it was dis-

84. Prunus subhirtella 'Pendula.' (Photo, Natl. Arborist Assoc.)

covered that various forms grow spontaneously in the wild in eastern Asia from Japan and Korea to central China.

Under long cultivation in Japan as the favorite flowering tree, *P. serrulata* has evolved scores of garden forms with variations in habit of growth and branching, flower size, flower color, and doubling of flowers. Many of these horticultural forms have, since the later part of the last century, also been introduced into the Western garden.

The other commonly planted species is *P. yedoensis,* a species of smaller stature. It is known in Japan only as a cultivated tree and is supposed to be a hybrid between *P. serrulata* and *P. subhirtella.* The latter is a small tree, growing to 30 feet in height and is much planted in Japan in its pendulous form (Wilson, 1916) (Figure 84).

12

Other Important Genera of
Shade Trees

The more important and prominent genera of shade trees are in general those with a large number of species widely distributed throughout the Northern Hemisphere. The species native to different areas are usually cultivated locally, sometimes since very ancient times. In the past three centuries, most of these species have been introduced into other continents and widely planted outside of their native locations. A few of these species have proven to be even more successful in their growth or more valuable in their use in certain foreign countries than in their own. Others, species of the same genus originally but of distinct and widely separate geographical ranges, have produced new hybrids with desirable characters when brought together in cultivation.

The more important genera of shade trees like *Acer, Aesculus, Gleditsia, Populus, Salix,* and *Tilia* have already been discussed. The following additional genera are treated in this chapter: *Alnus, Betula, Carpinus, Castanea, Fagus, Fraxinus, Liquidambar, Ostrya, Quercus,* and *Sorbus.* Other important genera of shade or ornamental trees of more limited geographical range are mentioned in chapters dealing with the various regions. References to the genera herein discussed may be found in such standard treatises as Elwes and Henry (1906-1913), Bailey (1933), Rehder (1940), and many other works of a general nature cited under the various chapters.

ALNUS, THE ALDERS

The alders, species of the genus *Alnus,* are not considered important as timber or ornamental trees, but are used as shade trees in certain situations. Two species have long been cultivated in Europe. The black alder, *A. glutinosa,* of wide range in Europe, Asia, and North Africa, has been introduced into eastern North America. It grows well especially in swampy areas, and occasionally escapes. The gray alder, *A. incana,* is a variable plant widely distributed in Europe, the Caucasus, as well as in North America. The form in the Old World is called variety *vulgaris,* a larger plant with many forms in cultivation. The Caucasian alder, *A. subcordata,* of Caucasia to Iran (Persia), is sometimes also cultivated.

BETULA, THE BIRCHES

Most species of the genus *Betula* are shrubs. Among the few species that attain tree size, the earliest cultivated is the European silver birch, *B. pendula,* a handsome tree with exfoliating white bark. Its natural range covers most of Europe and Asia Minor. A number of varieties are now in cultivation, the most commonly planted being 'Dalecarlica' ('Laciniata'), a highly ornamental tree with deeply cut leaves.

Another European species, *B. pubescens,* of northern and central Europe to Siberia, is less commonly planted.

CARPINUS, THE HORNBEAMS

The European hornbeam, *Carpinus betulus,* is widely distributed in Europe and it occurs also in Asia Minor, the Causasus, and western Asia. It is relatively unimportant as a timber tree, but it has long been planted as an ornamental tree and is used sometimes for making tall hedges. Under cultivation various pyramidal, fastigiate, pendulous, and cut and variegated-leaved forms have originated.

CASTANEA, THE CHESTNUTS

The Spanish or European chestnut, *Castanea satvia,* is an inhabitant of the mountain forests of the temperate regions of Europe, North Africa, and western Asia. It is a large tree with large edible nuts which are much esteemed as an important article of food in Spain, France, and Italy since early times. There are also cut-leaved and variegated-leaved varieties grown as ornamentals. The tree is very susceptible to the attack of the blight fungus that has destroyed the American chestnut, *C. dentata.* The eastern Asiatic species, the Chinese chestnut, *C. mollissima,* cultivated there since ancient times for its edible fruit, is more immune to the disease.

FAGUS, THE BEECHES

The common or European beech, *Fagus sylvatica,* is indigenous to England and northern, western, and southern Europe. It is an important timber tree and has long been cultivated there as an ornamental and shade tree.

There are some ten species in the genus *Fagus,* widely distributed in the temperate regions of the Northern Hemisphere, but the common beech is the only species that is widely and successfully cultivated. It is one of the very few European trees that thrive in eastern North America. This is an excellent avenue tree and is also widely used in parks as a specimen tree. In Europe it is commonly used for hedges.

Under cultivation there have developed numerous forms differing from each other in the shape and color of leaves and in habits. Some of these have originated as wild plants in the forests. The most familiar variety is the purple or copper beech, 'Atropunicea' or 'Purpurea,' a form with purplish leaves, known since 1680. All the trees in cultivation are supposed to have been derived from a single tree in the Hanleiter forests near Sondershausen, in Thuringia, central Germany. Propagation of this tree is by grafting or sometimes by seeds, but only a small percentage of the seedlings come purple.

Besides the purple beech, the other better-known varieties in cultivation include:

'Laciniata': fernleaf beech; with cut leaves. There are many other forms, similar to this variety and usually also called fernleaf beeches, such as 'Aspleniifolia,' with very narrow leaves; 'Rohanii,' with purple leaves; and others.

'Pendula': weeping beech; with drooping branches. This form has been observed in France in the wild. There is also a weeping form of the purple beech (Figure 85).

'Tortuosa': parasol beech; a dwarf form with twisted and contorted branches, drooping at the tips. This form has also been observed in the wild in France.

85. Fagus sylvatica 'Pendula.'

FRAXINUS, THE ASHES

Of the sixty-odd species of the genus, the European ash, *Fraxinus excelsior,* is the most widely planted. In its natural habitat, it is spread through almost all Europe to Asia Minor, being a timber tree valued for its toughness and elasticity. In Europe, it has long been cultivated as an ornamental tree.

The European ash has little variation in its natural populations, but numerous varieties have originated in cultivation. There are a number of variegated-leaved and very-narrow-leaved forms. Drooping and dwarf forms are also known. Perhaps the more remarkable is 'Diversifolia' (or (Monophylla'), usually called laurel-leaved or simple-leaved ash, with usually all simple or sometimes also with three-parted leaves. It is most distinct in appearance and is usually propagated by cuttings. It is sometimes met with in the wild state in the forests near Nancy in France and also more rarely in England and Ireland (Elwes & Henry, 1919).

In the southern part of Europe, two other ashes, the narrow-leaved ash, *F. angustifolia,* and the flowering ash, *F. ornus,* are also cultivated. Their natural ranges extend from southern Europe to western Asia.

LIQUIDAMBAR, THE SWEET GUMS

The sweet-gum genus, *Liquidambar,* has four species of disjunct ranges in the Northern Hemisphere: *L. styraciflua* of eastern North America, *L. macrophylla* of the mountains of Central America, *L. formosana* of eastern Asia, and *L. orientalis* of western Asia.

With the exception of the native sweet gum, *L. styraciflua,* the other species are little cultivated in this country as they are not as hardy as the American species. The American sweet gum was first cultivated in the later part of the seventeenth century, but the two Old World species have been in cultivation in their native regions since very ancient times. The eastern Asiatic species is much valued there as an ornamental tree, especially for its brilliant autumn foliage. Together with the maples, these trees are prominent in Oriental literature and art. The western Asiatic species is known for the fragrant resin exuding from the bark, which gives the name to the genus from a combination of the Latin for liquid and Arabic for amber.

OSTRYA, THE HOP HORNBEAMS

There are about seven species of *Ostrya* in the Northern

Hemisphere, in America, Europe, and Asia. The species are small or medium-sized trees, used locally as ornamental or shade trees, such as the European hop hornbeam, *O. carpinifolia*, in southern Europe and Asia Minor; the American hop hornbeam, *O. virginiana*, in eastern North America; and the Japanese hop hornbeam, *O. japonica*, in Japan, northeastern Asia, and China. These three species are so similar to each other in appearance and general features that some botanists consider them as geographical variants of the same species.

QUERCUS, THE OAKS

Oaks are among the most majestic trees, and the native species are used everywhere as street or shade trees. In most parts of Europe, two species have long been cultivated, the English oak, *Quercus robur,* native of Europe, North Africa, and western Asia, and the durmast oak, *Q. petraea,* native of Europe and western Asia. Both species are very large trees and are quite variable, with many ornamental forms in cultivation, but the English oak is the more commonly planted.

In the southern part of Europe, there are additional, less hardy species in cultivation. These include evergreens like the holm oak, *Q. ilex,* of southern Europe, and the cork oak, *Q. suber,* of southern Europe and North Africa, and deciduous trees like *Q. frainetto* of the Balkan Peninsula and southern Italy. Other species are found in southern Europe, extending to western Asia. Many species from North America and eastern Asia are also occasionally planted in other parts of the world. In eastern North America the native species are mostly planted as specimen or lawn trees, with the possible exception of the pin oak, *Q. palustris,* which is sometimes used for avenues and streets.

SORBUS, THE MOUNTAIN ASHES

There are more than eighty species in the genus *Sorbus*, distributed throughout the Northern Hemisphere, in Europe, America, and eastern Asia. These species are usually shrubs or shrublike small trees, while the few European species are generally of larger stature, attaining to a height of 40 feet or sometimes even to 60 feet. These trees, several of them long cultivated, are quite ornamental, especially with their showy bright-red fruits. The Rowan tree, or European mountain ash, *S. aucuparia,* of Europe to western Asia, has many cultivated varieties. It has been introduced into North America and naturalized in some areas. *S. domestica,* the service tree, of southern Europe, North Africa, and western Europe, is sometimes also grown for its fruit, which is used in France and Germany for cidermaking. Other cultivated species are the checker tree, *S. torminalis,* and the white beam tree, *S. aria.*

13

Cultivated Shade Trees Originated in Europe and Western to Central Asia

The flora of Europe is relatively impoverished as compared with some other parts of the temperate world such as eastern Asia and eastern North America. The number of tree species is not great and practically all of them have been in cultivation since ancient times for use as ornamental or shade trees or for some economic purpose (Loudon, 1838; Schneider, 1904-12; Elwes & Henry, 1906-13; Bean, 1950).

Most of the tree species of Europe are wide-ranging in their distribution, not only throughout the European continent but often extending also to western Asia and sometimes even to central and eastern Asia. Some of the southern European species may also extend to northern Africa. Few tree species are confined to Europe only. The cultivated trees of those wide-ranging species in other parts of the temperate world are likely to have been originated from European populations, as those from western Asia and northern Africa are generally not adapted to strictly temperate climates.

Because of the success of Europeans in colonizing different parts of the world in the last four centuries, European trees have, since Columbus, been widely dispersed to various parts of the world. These trees are not particularly successful in other temperate regions where ecological conditions are widely different. The most successful cultivated trees are those of hybrid or cultivated origin, such as the London plane and the Lombardy poplar, which are among the most widely planted trees of the world.

European trees have been introduced into North American since colonial times. All European tree genera occur also in eastern North America and therefore these European species all have their native counterparts. These introduced species usually do not thrive as well as the native ones, although a few, such as European beech and Norway maple, are among the most successful cultivated trees in this part of the country. Those trees originated in the Mediterranean region are less hardy and can be grown only in milder areas like California and the southern states (Rehder, 1940).

Western to central Asia, covering the region generally referred to as the Near East, is the cradle of human civilization. This area includes Asia Minor, Caucasus, Iraq, and Iran (Persia). This is the center of origin of wheat, rye, grape, cherry, almond, pear, fig, walnut, pomegranate, and many other important food crops.

As mentioned before the natural ranges of most of the trees native to Europe extend also to this region or at least to western Asia. The flora of this region is not much richer than that of Europe. Besides those wide-ranging species there are also others with their range primarily in western and central Asia but usually extending also to the southeastern corner of Europe(though not to the other parts of Europe).

Many trees of this area are among the earliest cultivated plants of mankind. However, these trees are not widely planted in other parts of the world because they are not as hardy as other temperate species and are generally suitable only to warm temperate or subtropical climates.

The following lists enumerate separately shade trees originated in Europe, in Europe to western and central Asia, and in western and central Asia. The numerous fruit trees are not here included, with the exception of a few such as the walnut and pomergranate, which are often also used as shade trees.

I. Important shade trees originated in Europe only.

Acer platanoides	Norway maple	Europe, Caucasia
Aesculus hippocastanum	Common horse chestnut	Balkan penin.
Fagus sylvatica	European beech	C. & S. Europe
Laurus nobilis	Laurel	Mediterranean region
Quercus frainetto	Italian oak	Balkan penin., Italy
Q. ilex	Holm oak	S. Europe
Q. lusitanica	Lusitanian oak	Spain, Portugal
Sorbus aria	White beam tree	Europe
Tilia cordata	Small-leaved linden	Europe
T. platyphyllos	Large-leaved linden	Europe
Ulmus procera	English elm	England, W. & C. Europe

II. Important shade trees originated in Europe to western and/or central Asia (sometimes also to North Africa).

Acer campestre	Hedge maple	Europe, W. Asia
A. pseudo-platanus	Sycamore maple	Europe, W. Asia
A. tataricum	Tatarian maple	Europe, W. Asia
Alnus glutinosa	Black alnus	Europe, Asia Minor
A. incana	Speckled alnus	Europe, N. Africa to C. Asia
Betula pendula	European birch	Europe, Asia Minor

Buxus semper-virens	Box	S. Europe, N. Africa, W. Asia
Carpinus betulus	European hornbeam	Europe to C. Asia
Castanea sativa	Spanish chestnut	S. Europe, N. Africa, W. Asia
Cornus mas	Cornelian cherry	C. & S. Europe, W. Asia
Fraxinus angustifolia	Narrow-leaved ash	S. Europe, N. Africa, W. Asia
F. excelsior	European ash	Europe, Asia Minor
F. ornus	Flowering ash	S. Europe, W. Asia
Ilex aquifolium	English holly	S. Europe, N. Africa, W. to E. Asia
Ostrya carpinifolia	European hop hornbeam	S. Europe, Asia Minor
Populus alba	White poplar	C. Europe to C. Asia
P. canescens	Gray poplar	Europe, W. Asia
P. nigra	Black poplar	Europe, W. Asia
P. tremula	European aspen	Europe, N. Africa, W. Asia
Quercus cerris	Turkey oak	S. E. Europe, W. Asia
Q. petraea	Durmast oak	Europe, W. Asia
Q. pubescens	Pubescent oak	S. Europe, Caucasia, W. Asia
Q. robur	English oak	Europe, N. Africa, W. Asia
Q. suber	Cork oak	S. Europe, N. Africa
Salix alba	White willow	Europe, N. Africa to C. Asia
S. amygdalina	Almond-leaved willow	Europe, W. to E. Asia
S. caprea	Goat willow	Europe to C. & N. E. Asia
S. cinerea	Gray willow	Europe to C. Asia
S. fragilis	Crack willow	Europe, W. Asia
S. pentandra	Bay willow	Europe, Caucasia

S. purpurea	Purple osier	Europe, N. Africa to E. Asia
S. viminalis	Common osier	Europe to N. E. Asia
Sorbus aucuparia	European mountain ash	Europe to W. Asia
S. domestica	Service tree	S. Europe, N. Africa, Asia Minor
S. torminalis	Checker tree	Europe, N. Africa, Asia Minor
Tilia dasystyla	Caucasian linden	S. E. Europe, Caucasia, N. Iran
T. petiolaris	Pendent silver linden	S. E. Europe, W. Asia
T. tomentosa	Silver linden	S. E. Europe, Caucasia, W. Asia
Ulmus carpinifolia	Smooth-leaved elm	Europe, N. Africa, W. Asia
U. glabra	Wych elm	N. & C. Europe to W. Asia
U. laevis	European white elm	C. Europe to W. Asia

III. Important shade trees originated in Caucasia, western and/or central Asia only.

Acer cappadocicum	Coliseum maple	Caucasia, W. Asia to Himalayas
A. velutinum	Velvet maple	Caucasia, N. Iran
Alnus subcordata	Caucasian alder	Caucasia
Fagus orientalis	Oriental Beech	Asia Minor, Caucasia, Iran
Liquidambar orientalis	Oriental sweet gum	W. Asia
Morus nigra	Black mulberry	W. Asia
Platanus orientalis	Oriental plane	W. Asia
Quercus libanii	Lebanon oak	Asia Minor
Zelkova carpinifolia	Zelkova elm	Caucasia

ENDEMIC GENERA

With the exception of small shrubby plants, few genera of woody plants are endemic to Europe, western Asia, and central Asia. Among those that nearly approaching tree size are the genera *Mespilus, Parrotia, Parrotiopsis, Punica,* and *Laurus.*

Mespilus, with its only species, *M. germanica,* the medlar, occurs from southeastern Europe to Iran. It is a shrub or small tree long cultivated in central Europe and England but probably not native there, and is sometimes grown for its fruit. *Parrotia* and *Parrotiopsis* are two monotypic genera of the Hamamelidaceae of Iran and the Himalayas respectively. They are small trees scarcely cultivated either locally or in other countries.

Punica granatum, the pomegranate, native to a belt extending from southeastern Europe to the Himalayas, is one of the most well-known and widely planted trees originated in this area. It was first cultivated probably in the Persian Gulf region. It is, however, a small tree or shrub, planted more as a fruit tree or for its ornamental flowers rather than as a shade tree.

Laurus is a genus of two species endemic to the Mediterranean region. The laurel, or sweet bay tree, *L. nobilis,* is the most widespread tub plant. It has been cultivated in Europe since time immemorial and is well known in history and poetry. The leaves were made into wreaths to crown heroes in ancient times. It is an evergreen which sometimes attains the height of 40 to 60 feet but rarely assumes a true treelike form. In cultivation, it is usually grown in trained or sheared forms of standard, globular, oval, conical, and pyramidal shapes for the formal types of gardens. The plant endures neglect and withstands repeated trimming year after year. It is thus most popular for outdoor decorations in both Europe and America.

14

Eastern Asia as a Center of Origin of Cultivated Trees

The richness of the flora of eastern Asia is well known in plant geography. Not only are the species and genera very numerous, but woody plants—trees as well as shrubs—are especially abundant. The concentration of woody plants in eastern Asia is probably the greatest outside of the tropics. Sargent (1894) estimated that in Japan proper there are not less than 550 species of woody plants or 1 in every 4.55 of the whole flora. No up-to-date census of the eastern Asiatic flora as a whole is available, but the known number of ligneous species as well as the proportion of woody plants may be even greater than that of Japan alone. In China, no less than 959 genera of woody plants were reported in 1935 (Hu, 1935). This exceeds the number of woody genera of all of the rest of the North Temperate Zone and is more than three times the number (313) found in eastern North America.

This richness of the woody flora of eastern Asia, especially mainland China, is due to its great diversity in topographic, climatic, and ecologic conditions. Historically, the absence of extensive glaciation during the Pleistocene made possible the preservation of a large number of genera formerly extensively distributed but later made extinct in most other parts of the world. Geographically the lack of impassable physiographic barriers between the temperate and tropical regions permits the mingling of numerous tropical elements with the temperate ones. Such a mixing of distinctly tropical and temperate plant groups occurs only in a few places in the whole world but never on such an extensive scale as in eastern Asia (Li, 1953).

Thus in eastern Asia, nearly all the important tree genera of the North Temperate Zone are represented. The only important genus not found there is *Robinia*, which is endemic to North America. There are widely ranging genera like the willows, poplars, oaks, elms, beeches, birches, hazels, etc., and others that are of limited distribution. Among the genera of narrow ranges, some are endemic only to eastern Asia, as those enumerated and discussed below. Others occur disjunctly also in western Asia like *Pterocarya* and *Zelkova*, or in eastern North America such as *Liriodendron, Halesia, Gymnocladus, Carya, Cladrastis, Sassafras, Catalpa, Magnolia,* and *Nyssa. Liquidambar* and *Platanus* are present disjunctly in all three areas (Li, 1952).

The presence of a large number of genera primarily of tropical distribution in temperate eastern Asia is a distinct feature of the flora. Usually only one or a few species of these otherwise large genera are found in eastern Asia, representing the northernmost attenuation of their ranges. The genera of trees so represented include *Ailanthus, Aphananthe, Broussonetia, Cedrela, Cinnamomum, Cudrania, Evodia, Firmiana, Lagerstroemia, Melia, Phoebe, Picrasma, Pistacia, Sapindus,* and *Sapium.* Some of the species are adaptable only to subtropical or warm temperate climates, but others, for instance those of *Ailanthus, Aphananthe, Broussonetia, Cedrela,* and *Evodia,* are perfectly hardy in all temperate regions.

With this large number of endemic genera and species, it is no wonder that eastern Asia becomes the major center of origin for horticultural plants. More Asiatic plants are now in cultivation in European and American gardens than native plants (Smith, 1931; Merrill, 1933; Rehder, 1935; Fogg, 1942). Among these introductions are a large number of trees and shrubs cultivated in China and Japan since earliest times. Others were introduced into cultivation in the last century by botanical explorers from the more remote hinterlands. Many of these are fast becoming popular garden subjects in the West either because of their novelty, distinct ornamental features, or other merits.

The following lists contains the more important and popular trees, ancient as well as recent in cultivation. For general information on Asiatic trees, the works of Chen (1936), Chun (1922), Kudo (1930), Rehder (1940), Shirasawa (1941), Sargent (1894), and others can be consulted.

The more important cultivated shade trees originated in eastern Asia.

Acer buergerianum	Trident maple	E. China, Japan
A. japonicum	Full-moon maple	Japan
A. mandshuricum	Manchurian maple	Manchuria, Korea
A. mono	Mono maple	China, Korea
A. truncatum	Purple-blow maple	N. China
Aesculus chinensis	Chinese horse chestnut	N. China
A. turbinata	Japanese horse chestnut	Japan
Ailanthus altissima	Tree of Heaven	China

Alangium platani-folium	Plain-leaved alangium	C. China to Japan
Albizzia julibrissin	Silk tree	C. China to Iran
A. kalkora	Lebbek albizzia	C. China to Japan
Alnus japonica	Japanese alnus	N. E. Asia, Japan
Aphananthe aspera	Aphananthe	E. China, Korea, Japan
Betula chinensis	Chinese birch	N. E. Asia, N. China
B. maximowiczii	Monarch birch	Japan
Broussonetia papyrifera	Paper mulberry	China, Japan
Carpinus cordata	Heart-leaved hornbeam	China, Japan
C. japonica	Japanese horn-beam	Japan
Castanea crenata	Japanese chestnut	Japan
C. mollissima	Chinese chestnut	China, Korea
Catalpa bungei	Manchurian catalpa	N. China
C. ovata	Chinese catalpa	China
Cedrela sinensis	Chinese cedrela	China
Celtis bungeana	Bunge hackberry	N. E. Asia
C. koreana	Korean hackberry	N. E. Asia
Cercidiphyllum japonicum	Katsura tree	China, Japan
Cinnamomum camphora (Figure 86)	Camphor tree	S. China
Cladrastis sinensis	Chinese yellow-wood	W. & C. China
Cornus kousa	Kousa dogwood	Korea, Japan
C. officinalis	Japanese cornel	Korea, Japan
Corylus chinensis	Chinese hazel	W. & C. China
Cudrania tri-cuspidata	Cudrania	China, Korea, Japan
Davidia involu-crata	Dove tree	W. China

86. *Cinnamomum camphora (Photo, U.S. Forest Service.)*

Diospyros kaki	Kaki persimmon	China, Japan
D. lotus	Date-plum persimmon	W. Asia to China, Japan
Dipteronia sinensis	Dipteronia	C. China
Emmenopterys henryi	Emmenopterys	China
Eucommia ulmoides (Figure 87)	Chinese rubber tree	C. China
Evodia daniellii	Beebee tree	N. China, Korea
E. officinalis	Chinese evodia	C. & W. China
Fagus japonica	Japanese beech	China, Japan

87. *Eucommia ulmoides.*

Firmiana simplex	Phoenix tree	China
Fraxinus bungeana	Bunge ash	N. China
F. chinensis	Chinese ash	China
F. mandshurica	Manchurian ash	N. E. Asia
Gleditsia sinensis	Chinese honey locust	E. China
Gymnocladus chinensis	Chinese coffee tree	C. China
Hovenia dulcis	Raisin tree	China
Idesia polycarpa	Idesia	China, Japan
Juglans mandshurica	Manchurian walnut	N. E. Asia
Kalopanax pictus	Kalopanax	China, Japan
Koelreuteria paniculata	**Goldenrain tree**	China, Korea, Japan
Lagerstroemia indica	Crape myrtle	China

Liquidambar formosana	Formosa sweet gum	W. to E. China
Maackia amurensis	Amur maackia	Manchuria, N. E. Asia
Magnolia denudata	Yulan magnolia	C. China
M. liliflora	Lily magnolia	China
Melia azedarach	China tree	S. China to Himalayas
Morus alba (Figure 88)	White mulberry	China

88. Morus alba.

M. australis	Japanese mulberry	China, Korea, Japan
Nyssa sinensis	Chinese tupelo	C. China
Ostrya japonica	Japanese hop hornbeam	China, Japan
Paulownia tomentosa	Paulownia	China
Phellodendron amurense	Amur cork tree	N. China, N. E. Asia
Phoebe nanmu	Nanmu	W. China
Picrasma quassioides	Quassia wood	Himalayas, China, Japan
Pistacia chinensis	Chinese pistachio	China
Poliothyrsis sinensis	Pearl-bloom tree	C. China
Populus adenopoda	Chinese aspen	C. & W. China
P. maximowiczii	Japanese poplar	N. E. Asia, Japan
P. sieboldii	Japanese aspen	Japan
P. simonii	Simon's poplar	N. China
P. tomentosa	Chinese white poplar	N. China
Pterocarya stenoptera	Chinese wing nut	China
Pteroceltis tatarinowii	Wing celtis	N. & C. China
Pterostyrax hispida	Epaulette tree	Japan
Quercus actutissima	Saw-tooth oak	Himalayas to Japan
Q. aliena	Oriental white oak	C. China, Japan
Q. dentata	Daimyo oak	N. E. Asia, N. Japan
Q. mongolica	Mongolian oak	Himalayas to Japan
Q. variabilis	Oriental oak	N. China, Korea, Japan

Salix babylonica	Weeping willow	China
S. matsudana	Hankow willow	China
Sapindus mukor-ossi	Chinese Soap-berry	Himalayas to E. Asia
Sapium sebiferum	Chinese tallow tree	China
Sassafras tsumu	Chinese sassafras	C. China
Sophora japonica (Figure 89)	Chinese scholar tree	China, Korea
Styrax japonica	Japanese snow-bell	China, Japan
S. obassia	Fragrant snowbell	Japan
Symplocos pani-culata	Asiatic sweet leaf	Himalayas, China, Japan
Tapiscia sinensis	False pistachio	C. China
Tetracentron sinense	Tetracentron	C. & W. China

89. Sophora japonica.

Tilia japonica	Japanese linden	Japan
T. mandshurica	Manchurian linden	N. E. Asia
T. miqueliana	Miquel linden	E. China
T. mongolica	Mongolian linden	China, Mongolia
T. tuan	Tuan linden	C. China
Ulmus japonica	Japanese elm	N. E. Asia, Japan
U. parvifolia (Figure 90)	Chinese elm	N. & C. China, Korea, Japan
U. pumila	Siberian elm	E. Siberia, N. China
Zelkova serrata	Japanese zelkova	Japan

90. *Ulmus parvifolia.*

ENDEMIC GENERA

The very large number of endemic genera of woody plants in eastern Asia contain numerous useful and interesting trees, many of which are now used extensively all over the world as shade trees. Twelve endemic genera are described herein. To these may be added the genera *Davidia, Koelreuteria,* and *Paulownia* which have been discussed elsewhere. While most of these are valued as shade trees, some are planted mainly because of their flowers and a few are planted as novelties. These latter are included to show the rich and varied nature of the flora.

CERCIDIPHYLLUM

Cercidiphyllum, or katsura tree, is one of the most impressive trees of the temperate world, attaining with age an immense size, surpassing that of most other broadleaved trees. Botanically it is interesting as it is the only genus of the family Cercidiphyllaceae, one of several small relic families of flowering plants surviving only in eastern Asia. It is a tree of graceful habit and handsome foliage, with the heart-shaped leaves resembling those of *Cercis,* the Judas tree, hence the name meaning "cercis leaf." *C. japonicum* is a forest tree of Japan. Several forms with smaller or larger leaves are sometimes distinguished as separate species. It was first introduced into America in about 1878. There is a Chinese variety, *sinense,* occurring disjunctly in the mountains of western China and which was first introduced into cultivation by E. H. Wilson in about 1907.

DIPTERONIA

Dipteronia is another unique tree genus discovered in

China. It is the only other genus of the Aceraceae besides the maple genus *Acer,* differing from the latter mainly in the seed being winged all around. The leaves are pinnately compound. There are two species in the genus, one of which, *D. sinensis,* has been introduced into cultivation. It was first introduced by A. Henry in 1900 from western Hupeh.

EMMENOPTERYS

Emmenopterys, with its only species, *E. henryi,* is a tree widespread in the western part of China. It has been described as one of the most beautiful trees of the Chinese forests because of its showy flowers with their large bracts. The bracts are white at first, turning pink later and persisting on the fruit. The tree was first introduced into cultivation by E. H. Wilson in 1907 from western Hupeh.

EUCOMMIA

Eucommia ulmoides (Figure 87), the Chinese rubber tree, is the only species of the genus, which is in turn the monotype of the Eucommiaceae, another relic family found only in China. It is a tree of moderate size resembling somewhat the elm in general appearance and leaf shape. It is handsome, with dense, dark-green foliage. The tree, because of the presence of latex in the leaves and bark, is of special interest as the only rubber-producing tree hardy in temperate regions. The rubber, however, is not present in sufficient quantity for commercial usage.

This tree has long been cultivated in central and western China since very early times primarily for its bark, which is used as a drug. All the existing trees seem to be cultivated and no truly wild trees have been noted. It was first introduced into Europe and America toward the end of the last century.

HOVENIA

Hovenia, with its only species, *H. dulcis,* is native to China and has long been cultivated in that country. From China, its cultivation has also been extended early to India and Japan. It is planted mainly for its handsome foliage, which resembles somewhat that of the linden. The fleshy fruiting stalks are sweet and edible, hence the name raisin tree. It was first introduced into the West in 1920 by A. B. Lambert in England, who raised it from seeds brought from China by Staunton (Bretschneider, 1898).

IDESIA

The family Flacourtiaceae is primarily a tropical one but there are several genera endemic to temperate eastern Asia. The most well known of these genera is *Idesia,* a monotypic genus of Japan and central and western China. There is only one species, *I. polycarpa*, a medium-sized tree long cultivated in eastern Asia as a shade or ornamental tree. The leaves are said to be edible. The most distinct feature is the large panicles of small, yellow-green flowers. It was introduced into Europe by the Russian botanist Maximowicz in 1864 from Japan (Bretschneider, 1898).

In temperate China, there are two other genera of the family Flacourtiaceae. *Poliothyrsis* is a monotypic genus of central China. The species, *P. sinensis,* is a slender tree. It was first discovered in the mountains of western Hupeh in 1889 by A. Henry and introduced into cultivation by E. H. Wilson in 1908. *Carrierea* has two or three species in southwestern China, the most well known being *C. calycina,* a handsome small- or medium-sized tree with panicles of white flowers. It was first discovered by Père Paul Farges in 1896 and introduced into cultivation by E. H. Wilson from western Szechuan. Both *Poliothyrsis* and *Carrierea* are still very rare in cultivation.

KALOPANAX

Kalopanax is a monotypic genus of the Araliaceae, the ginseng family. *K. pictus,* a tall, prickly tree with stout branches and large, palmately lobed leaves, is very distinct with its tropical appearance. It is native to eastern Asia (in China, Korea, and Japan), as a tree of the forests and is also occasionally planted. It was introduced into Europe from Japan around 1865 by Maximowicz.

MAACKIA

There are two or three species in *Maackia,* a genus closely related to *Cladrastis,* the yellowwood. The common species is *Maackia amurensis,* a species native to northeastern Asia and Japan, and also planted there. It is a medium-sized tree, very similar to *Cladrastis* in general appearance except that the leaflets are opposite or subopposite in arrangement. Maximowicz introduced the tree to St. Petersburg in 1875 from the Amur region (Bretschneider, 1898).

PHELLODENDRON

Several species of *Phellodendron* are now in cultivation, the most well known being *P. amurense,* the Amur cork tree, of northeastern Asia to northern China and Japan. It is so named because of the thick corky bark. The inner bark, yellowish in color, is used in China and Japan as a drug. Maximowicz introduced the tree into Europe around 1856.

PTEROCELTIS

Pteroceltis tatarinowii, the only species of the genus, is native to northern and central China. It is related to the hackberry and resembles the latter closely, differing mainly in the winged fruits. It was introduced into Western gardens toward the end of the nineteenth century.

TETRACENTRON

Tetracentron is a monotypic genus native to central and western China, one of the relic woody types preserved in that area. The species *T. sinense* is similar to *Cercidiphyllum* in appearance but can be distinguished by the alternate leaves. It was first named in 1889 and later introduced into America by E. H. Wilson in 1908, but it is still rare in cultivation.

15

North America as a Center of Origin of Cultivated Trees

The flora of North America is a rich one. Many genera and species of trees are represented. These trees, however, came under human cultivation much later than those of Europe and Asia, as few were ever planted by the aboriginal Indians. The early colonists introduced and planted many trees from the Old World but paid relatively little attention to the native American species. A number were brought under cultivation around the middle of the eighteenth century, and gradually more and more native trees were planted. The majority of these native species were first introduced into cultivation in the nineteenth century and now practically every species of trees is in cultivation one way or the other.

Most of these American species are planted or used more or less locally but a few, such as the black locust and the honey locust, have become widespread also in other continents. Many North American plants were first introduced into cultivation in Europe. The literature on the introduction of North American plants into Europe is very voluminous. For more recent studies on the subject the works of Wein (1930-1932), Wood and Mathews (1957),

Leroy (1957), Galoux (1957), and Guillaumin and Chaudun (1957) may be consulted.

There are two major centers of origin of American trees, the eastern part of the continent and the Pacific coastal region. The floras of these two regions are widely different in nature. Eastern North America has not had any pronounced geological disturbances since the beginning of the Tertiary some sixty million years ago, and hence the flora is more stable and archaic. Here are preserved many genera and species which have since disappeared in other portions of the world because of glaciation, submergence, or other geological and climatic changes. Another great area of similar nature is eastern Asia, and this is why there are so many common types in the vegetation of these two distant regions (Li, 1952).

On the other hand, the western part of North America has encountered considerable geological disturbances such as submergence, uplift, climatic shift, volcanic outflow, and some glaciation. As a result, the flora is entirely different from that of the East. There are many modern and highly specialized plant families and genera adapted to soils and habitats of more recent origin. The Rocky Mountains, cutting off moisture from the Pacific, made the region to the east a large, arid, treeless area that further separates the floras of the East and the West.

The trees from the Pacific states prefer a climate with mild winters and cool summers. They grow on soils that are of modern origin. These trees generally do not thrive in the eastern part of the continent but may be successfully grown in Europe. On the other hand, trees from eastern Asia are more suitable to eastern North America as the climate, soils, and geological history of these two distant regions are more similar (Fogg, 1942).

Some American tree species are of wide ranges, extending from the Pacific to the Atlantic coasts, especially in the northern part of the continent, but these are generally differentiable into various geological varieties or subspecies. Sometimes these wide-ranging species occur further south to the middle part of the continent. The great majority of the species are concentrated, as mentioned above, either in the East or in the coastal states of the West. The following lists of trees are grouped into these three units. These lists do not include many small trees planted mainly for their flowers or their fruits, such as species in the genera *Prunus, Amelanchier, Cercis, Kalmia,* and many others. There are mainly used as ornamentals and rarely as shade trees.

General information on American trees can be obtained from such excellent treatises as those by Marshall (1783), Michaux (1817-1819), Browne (1832), Nuttall (1842-1849), Sargent (1891-1902, 1926), Britton and Schafer (1908), Rehder (1940), Little (1949), and many others.

I. Important shade trees originated in North America, wide-ranging from east to west, or in the middle part of the continent.

Betula papyrifera	Canoe birch	Canada, N. U.S.A.
Carya pecan	Pecan	Mexico to C. U.S.A.
Populus tacama-haca	Balsam poplar	Alaska, Canada, N. U.S.A.
P. tremuloides	Quaking aspen	Alaska, Canada, to N. Mexico
Salix nigra	Black willow	Canada, U.S.A.

II. Important shade trees originated in eastern North America.

Acer negundo	Box elder
A. pseudo-platanus	Sycamore maple
A. rubrum	Red maple
A. saccharinum	Silver maple
A. saccharum	Sugar maple
Aesculus octandra	Sweet buckeye
A. pavia	Red buckeye

Betula lenta	Cherry birch
B. lutea	Yellow birch
B. nigra	River birch
Carpinus caroliniana	American hornbeam
Carya cordiformis	Bitternut
C. ovata	Shagbark hickory
Castanea dentata	American chestnut
Catalpa bignonioides	Common catalpa
C. speciosa	Western catalpa
Cladrastis lutea	American yellowwood
Cornus florida	Flowering dogwood
Diospyros virginiana	Common persimmon
Fagus grandifolia	American beech
Fraxinus americana	White ash
F. nigra	Black ash
F. pennsylvanica	Red ash
F. tomentosa	Pumpkin ash
Gleditsia triacanthos	Honey locust
Gymnocladus dioicus	Kentucky coffee tree
Halesia diptera	Towing silver-bell tree
H. monticola	Mountain silver-bell tree
Juglans cinerea	Butternut
J. nigra	Black walnut
Liquidambar styraciflua	Blistered sweet gum
Maclura pomifera	Osage orange
Magnolia acuminata	Cucumber tree
M. grandiflora	Bull bay
M. macrophylla	Large-leaved cucumber tree
M. tripetala	Umbrella magnolia
Morus rubra	Red mulberry
Nyssa sylvatica	Black tupelo
Ostrya virginiana	American hop hornbeam
Oxydendrum arboreum	Sorrel tree
Platanus occidentalis	American plane
Populus deltoides	Cottonwood
P. grandidentata	Large-toothed aspen

Ptelea trifoliata	Hop tree
Quercus bicolor	Swamp white oak
Q. borealis	Red oak
Q. coccinea	Scarlet oak
Q. falcata	Spanish oak
Q. imbricaria	Shingle oak
Q. laurifolia	Laurel oak
Q. lyrata	Overcup oak
Q. macrocarpa	Bur oak
Q. marilandica	Blackjack oak
Q. montana	Chestnut oak
Q. nigra	Water oak
Q. palustris	Pin oak
Q. prinus	Basket oak
Q. stellata	Post oak
Q. velutina	Black oak
Q. virginiana	Live oak
Robinia pseudoacacia	Black locust
Salix discolor	Pussy willow
Sassafras albidum	Common sassafras
Tilia americana	American linden
T. heterophylla	Bee-tree linden
Ulmus americana	White elm
U. fulva	Slippery elm

III. Shade trees originated in western North American and sometimes also cultivated in other regions.

Acer macrophyllum	Oregon maple
Aesculus californica	California buckeye
Castanopsis chrysophylla	Giant evergreen chinkapin
Cornus nuttallii	Pacific dogwood
Fraxinus oregona	Oregon ash
F. velutina	Velvet ash
Lithocarpus densiflorus	Tanbark oak
Platanus racemosa	California plane tree
P. wrightii	Arizona plane tree
Populus trichocarpa	Western balsam poplar

Quercus chrysolepis	California live oak
Q. garryana	Oregon oak
Q. kelloggii	California black oak
Umbellularia californica	California laurel

ENDEMIC GENERA

The flora of North America, as mentioned before, is a rich one and there are numerous endemic genera of woody plants both in the East as well as the West. However, most of these are shrubby or at most small trees, rarely usable as street or shade trees. There are only a few of these endemic genera that attain truly tree size, such as *Maclura, Oxydendrum,* and *Robinia* of the East and *Umbellularia* of the West.

Robinia, one of the most important genera of cultivated trees, has been discussed in detail elsewhere. *Maclura,* with its only species *M. pomifera,* the Osage orange, is native from Arkansas to Oklahoma and Texas. It is a thorny tree that grows to a size of sixty feet in height and often used as a hedge plant in the Midwest. *Oxydendrum,* also with one species, *O. arboreum,* the sorrel tree, is native to the eastern United States, south from Pennsylvania and Indiana. It is also a medium-sized tree but sometimes attains a height of seventy-five feet. According to Rehder (1940) the Osage orange was first cultivated in 1818 and the sorrel tree in 1747.

Umbellularia, with its only species *U. californica,* the Californian laurel, is native to the West Coast from Oregon to California. It is named the California laurel because of its evergreen habit and its resemblance to laurel. It is a highly ornamental tree with dense lustrous foliage, but it is suitable only in milder regions. The tree has been in cultivation since 1829.

Bibliography

AITON, W. *Hortus Kewensis,* 2nd ed., 5 vols. London, 1810-1813.

ANDRÉ, E. *"Aesculus × plantierensis," Rev. hort.* (Paris, 1894), 246-247.

BAILEY, L. H. *The Standard Cyclopedia of Horticulture.* New ed. New York, 1950.

BARCLAY, J. G. "The name Ginkgo," *J. Roy. Hort. Soc.* (London), Vol. 69 (1944), 68-69.

BEAN, W. J. *Trees and Shrubs Hardy in the British Isles.* 7th ed. London, 1950.

———. "The London plane (*Platanus acerifolia*)," *Gardener's Chronicle,* Vol. III, No. 66 (1919), 47.

BERRY, E. W. *Tree Ancestors.* Baltimore, 1923.

BLÜMKE, S. "Beiträge zur Kenntnis der Robinie (*Robinia pseudoacacia* L.)," *Mitt. deut. dendrol. Ges.,* Vol 59 (1956), 38-65.

BRETSCHNEIDER, E. V. *Botanicon Sinicum.* 3 vols. Shanghai, 1882-1895.

———. *History of European Botanical Discoveries in China.* 2 vols. St. Petersburg, 1898.

BRITTON, N. L., and SCHAFER, J. A. *North American Trees.* New York, 1908.

BROWNE, D. J. *The Trees of America.* New York, 1857.

BURKILL, I. H. "On the dispersal of the plants most intimate to Buddhism," *J. Arnold Arboretum* (Harvard Univ.), Vol. 27 (1946), 327-339.

CHAÑEY, R. W. "Redwoods around the Pacific basin," *Pacific Discovery,* I, No. 5 (1948), 4-14.

CHEN, Y. *An Illustrated Manual of Forest Trees in China.* Nanking, 1936. In Chinese.

CHENG, W. C., and CHIEN, S. C. "An enumeration of vascular plants from Chekiang, I," *Contribs. Biol. Lab. Sci. Soc. China, Botan. Ser.*, Vol. 8 (1933), 298-306.

CHU, K., and COOPER, W. S. "An ecological reconnaissance in the native home of *Metasequoia glyptostroboides*," *Ecology*, Vol. 31 (1950), 260-278.

CHUN, W. Y. *Chinese Economic Trees.* Shanghai, 1922.

CROIZAT, L. "Oriental planes in New York City," *J. New York Botan. Garden*, Vol. 38 (1937), 62-64.

DANDY, J. E. "A survey of the genus *Magnolia* together with *Michelia* and *Mangelietia*," *In Camellias and Magnolia, Roy. Hort. Soc. Conf. Rep.* (1950), 64-81.

DECANDOLLE, A. P. *The Origins of Cultivated Plants.* London, 1898.

EAMES, A. J. "The seed and Ginkgo," *J. Arnold Arboretum*, Vol. 36 (1955), 165-170.

EDKINS, J. *Ancient Symbolism among the Chinese.* London and Shanghai, 1889.

ELWES, H. J., and HENRY, A. *Trees of Great Britain and Ireland.* 7 vols. Edinburgh, 1906-1913.

ENDLICHER, S. *Synopsis Coniferum.* Sangalli, 1847.

ENGLER, V. *Monographie der Gattung Tilia.* Berlin, 1909.

FAO. "Les Peupliers dans la production du bois et l'utilization des terres," *Coll. FAO*, No. 12 (1956).

FERÑALD, M. L. *Gray's Manual of Botany.* 8th ed. New York, 1950.

FLORIN, R. "On *Metasequoia*, living and fossil," *Botan. Notiser* (1952), 1-29.

FOCKE, W. O. "Das siechthum der Pyramidaenpappeln," *Garten-Zeit*, Vol. 2 (1883), 389-392.

FOGG, JR., J. M. "Eastern Asiatic plants in eastern American gardens," *Morris Arboretum Bull.*, Vol. 4 (1942), 15-20.

FORTUNE, R. *A Residence among the Chinese.* London, 1857.

———. *Yedo and Peking.* London, 1863.

FUJII, K. "On the nature and origin of the so-called 'chi-chi' (nipples) of *Ginkgo biloba* L.," *Botan. Mag.* (Tokyo), Vol. 9 (1895), 440-450.

————. "On the different views hitherto proposed regarding the morphology of the flowers of *Ginkgo biloba* L.," *Botan. Mag.* (Tokyo), Vol. 10, No. 2 (1896), 7-8, 13-15, 104-109.

GAGNEPAIN, F. "Un genre nouveau de Butomacées et quelques espèces nouvelles d'Indo-Chine," *Bull. soc. botan. France*, Vol. 86 (1939), 300-303.

GALOUX, A. "Les grandes étapes de l'introduction des arbres nord-americains en Belgique," *In Botanistes français en Amérique du Nord* (1957), 253-261.

GRESSITT, J. L."The California-Lingnan dawn-redwood expedition," *Proc. Calif. Acad. Sci.*, Vol. 27, No. 2 (1953), 25-58.

GUILLAUMIN, A. and CHAUDUN, V. "L'introduction en France des plantes horticoles originaires de l'Amérique du Nord avant 1850," *In Botanistes français en Amerique du Nord* (1957), 123-135.

HENRY, A., and FLOOD, M. C. "The history of the London plane, *Platanus acerifolia*, with notes on the genus *Platanus*," *Proc. Roy. Irish Acad.*, Vol. 35 (1919), 9-28.

HESMER, H. *Das Pappelbuch in Auftrage des Deutschen Pappelvereins*. Bonn, 1951.

HIRASE S. "Études sur la fécondation et l'embryogénie du *Ginkgo biloba*," *J. Coll. Sci. Imp. Univ. Tokyo*, Vol. 8 (1895), 307-322; Vol. 12 (1898), 103-149.

HJELMQVIST, H. "Studies on the floral morphology and phylogeny of the Amentiferae," *Botan. Notiser, Supp.*, Vol. 2 (1948), 1-117.

HOAR, C. S. "Chromosome studies in *Aesculus*," *Botan. Gaz.*, Vol. 84 (1927), 156-170.

HOOKER, W. J. "Occidental plane," *Gardener's Chronicle* (1856), 282.

Hu, H. H., and Cheng, W. C. "On the new family Metasequoiaceae and on *Metasequoia glyptostroboides*, a living species of the genus *Metasequoia* found in Szechuan and Hupeh," *Bull. Fan Mem. Inst. Biol.*, N. S. 1 (1948), 153-161.

Johnstone, G. *Asiatic Magnolias in Cultivation*. London, 1955.

Kaempfer, E. *Amoenitatum exoticarum*. Lemgoviae, 1712.

Koidzumi, G. "Old records of Ginkgo in China," *Acta Phytotax, Geobotan.*, Vol. 5 (1936), 263-264. In Japanese.

Kramer, S. N. *From the Tablets of Sumer*. Indian Hills, Colorado, 1956.

Kudo, Y. *Taxonomy of Useful Trees in Japan*. 2nd ed. Tokyo, 1930.

Leroy, J. F. "Note sur l'introduction des plantes américaines en France dans la première moitié du XVIIIe siecle," *In Botanistes français en Amerique du Nord* (1957), 285, 286.

Li, H. L. "Floristic relationships between eastern Asia and eastern North America," *Trans. Am. Phil. Soc.*, Vol. 42, No. 2 (1952), 371-429.

————. "Endemism in the ligneous flora of eastern Asia," *Proc. 7th Pacific Sci. Congr.*, Vol. 5 (1952), 212-216.

————. "The garden flowers of China," *Chron. Botan.*, No. 19 (1959).

Liñnaeus, C. *Mantissa plantarum*. 2 vols. 1767-1771.

Little, Jr., E. J. "Important forest trees of the United States," *U.S. Dept. Agric. Yearbook* (1949), 763-841.

"The Lombardy Poplar," *Gardner's Chronicle*, Vol. II, No. 20 (1883), 571-572.

Loudon, J. C. *Arboretum et fruticetum britannicum*. 8 vols. London, 1844.

Makino, T. *An Illustrated Flora of Japan*. Rev. ed. Tokyo, 1951. In Japanese.

Marshall, H. *Arbustum americanum*. Philadelphia, 1785.

Master, M. T. "Conifers. In Forbes & Hemsley, An enumeration of all plants known from China proper," *J. Linnean*

Soc. London Botany, Vol. 26 (1902), 540-559.

MATSUMURA, Y. "Maples of Japan," *Arboretum Bull.* (Seattle, Wash.), Vol. 17 (1954), 105-110.

MERRILL, E. D. "Eastern Asia as a source of ornamental plants," *N. Y. Botan. Garden,* Vol. 34 (1933), 238-243.

MICHAUX, F. A. *The North American Sylva.* 3 vols. Philadelphia, 1817-1819.

MIKI, S. "On the change of flora in eastern Asia since the Tertiary period," *J. Japan. Botan.,* Vol. 11 (1941), 237-303.

MILLER, P. *The Gardener's Dictionary,* 7th ed. London, 1759.

MOLDENKE, H. N., and MOLDENKE, A. L. *Plants of the Bible.* New York, 1952.

MOULE, A. C. "The name *Ginkgo biloba* and other names of the tree," *T'oung Pao,* Vol. 33 (1939), 193-219.

———. "The name Ginkgo," *J. Roy. Hort. Soc.,* Vol. 69 (1944), 166.

MULLIGAN, B. O. "Maples cultivated in the United States and Canada," *Am. Assoc. Botan. Gardens Arboretums* (1958).

NUTTALL, T. *The North American Sylva.* 3 vols. Philadelphia, 1842-1849.

PAULEY, S. C. "Forest tree genetics research: *Populus* L." *Econ. Botany,* Vol. 3 (1949), 299-330.

PULLE, A. "Over de Ginkgo alias Ginkyo," *Jaarb. Nederl. Dendr. Ver.* (1940-1946), 25-35.

REHDER, A. "On the history of the introduction of woody plants into North America," *Natl. Hort. Mag.,* Vol. 15 (1935), 245-257.

———. *A Manual of Cultivated Trees and Shrubs.* Rev. ed. New York, 1940.

———.*Bibliography of Cultivated Trees and Shrubs.* Jamaica Plain, Mass., 1949.

RIVERS, T. "The London plane trees," *Gardener's Chronicle,* Vol. 47 (1860), 255-257.

SARGENT, C. S. "The lindens of western Europe," *Garden & Forest,* Vol. 2 (1889), 255-257.

———. *The Silva of North America.* 14 vols. Boston and New York, 1891-1902.

————. *Forest Flora of Japan.* Boston and New York, 1894.

————. "Plantae wilsonianae," *Publ. Arnold Arboretum,* No. 4 (1911-1917).

SCHNEIDER, C. K. *Illustriertes Handbuch der Laubholzkunde.* 2 vols. Jena, 1904-1912.

SCHRAMM, J. R. "An experiment in greenhouse benches," *Morris Arboretum Bull.,* Vol. 4, No. 3 (1942), 25-26.

SCHREINER, E. J. "Production of poplar timber in Europe and its significance and application in the United States," *U.S. Dept. Agri. Tech. Bull.,* No. 150 (1959).

SEWARD, A. C. *Link with the Past in Plants.* Cambridge, 1911.

————. *Fossil Botany.* Cambridge, 1919.

————. *Plant Life Through the Ages.* Cambridge, 1931.

————. "The story of the maidenhair tree," *Sci. Progr.,* Vol. 32 (1938), 420-440.

————, and GOWAN, J. "The maidenhair tree (*Ginkgo biloba* L.), *Ann. Botany,* Vol. 14 (1900), 109-154.

SHAW, G. R. "A contribution to the anatomy of *Ginkgo biloba,*" *New Phytologist,* Vol. 7 (1908), 85-92.

SHIRASAWA, H. *Icones of the Forest Trees of Japan.* 2 vols. Tokyo, 1914. In Japanese.

SKIÑNER, H. T. "A lath house built for durability," *Morris Arboretum Bull.,* Vol. 4, No. 11 (1949), 91-93.

————."*Metasequoia* in its second year," *Morris Arboretum Bull.,* Vol. 4, No. 11 (1949), 94.

SMITH, J. E. "Characters of a new genus of plants named *Salisburia,*" *Proc. Linnean Soc.,* Vol 3 (1797), 330-332.

SMITH, J. R. *Tree Crops: a Permanent Agriculture.* New York, 1950.

SMITH, W. W. "The contribution of China to European gardens," *Notes Botan. Garden Edinburgh,* Vol. 16 (1931), 215-221.

SPRECHER, A. *Le Ginkgo biloba L.* Geneve, 1907.

SPRENGEL, K.•*Historia rei herbariae.* 2 vols. Amsterdam, 1808.

STANDLEY, P. C. "Trees and shrubs of Mexico (Fagaceae-Fabaceae)," *Contribs. U.S. Natl. Herb.,* Vol. 23, No. 2 (1922), 171-515.

STRASBURGER, E. *Die Coniferen und Gnetaceen.* Jena, 1872.

THOMMEN, E. "News zur Schreibung des Namens Ginkgo," *Verhandl. naturforsch. Ges.* (Basel), Vol. 60 (1949), 77-103.

TRUE, R. H. "Abstract of Dr. Lyle W. R. Jackson's lecture on diseases of the plane tree at the Morris Arboretum, January 11, 1936," *Morris Arboretum, Bull.,* Vol. 1 (1936), 22-23.

TSENG, M. C. "The *Ginkgo biloba* of Chuki District, Chekiang," *Hortus,* Vol. 1 (1935), 157-165. In Chinese.

UPCOTT, M. "The parents and progeny of *Aesculus carnea,*" *J. Genet.,* Vol. 33 (1936), 135-149.

VAVILOV, N. J. "Studies on the origin of cultivated plants," *Bull. Appl. Botany Plant Breeding* (Leningrad), Vol. 16, No. 2 (1926), 3-248. In Russian with English translation, pp. 130-248.

———."The origin, variation, immunity and breeding of cultivated plants," transl. by K. S. Chester. *Chron. Botan.,* No. 13 (1951).

WEIÑ, K. "Die erste Eifuehrung nordamericanischer Geholtze in Europe," *Mitt. deut. dendrol. Ges.,* Vol. 42 (1930), 137-163; Vol. 43 (1931), 95-154; Vol. 44 (1932), 123-129.

WILSON, E. H. "The cherries of Japan," *Publ. Arnold Arboretum,* No. 7 (1916).

———. "The conifers and taxads of Japan," *Publ. Arnold Arboretum,* No. 8 (1916).

———. *The Romance of our Trees.* New York, 1920.

WOOD, R. F., and MATTHEWS, J. D. "Arbres nord-américains et sylviculture anglaise," *In Botanistes française en Amérique du Nord* (1957), 303-313.

ZIRKLE, C. "The beginnings of plant hybridization," *Morris Arboretum Monograph,* No. 1 (1935).

Index To Scientific Names

Index To Common Names

Names in foreign languages are in italics.

Pennsylvania Paperbacks